1st International Conference on Advanced Research in Engineering Sciences and Technology (ICAREST 2021)

IOP Conference Series: Materials Science and Engineering
Volume 1184

Babylon, Iraq
6 – 8 April 2021

ISBN: 978-1-7138-4644-4
ISSN: 1757-8981

Printed from e-media with permission by:

Curran Associates, Inc.
57 Morehouse Lane
Red Hook, NY 12571

Some format issues inherent in the e-media version may also appear in this print version.

Institute of Physics
Dirac House, Temple Back
Bristol BS1 6BE UK

Phone: 44 1 17 929 7481
Fax: 44 1 17 920 0979

techtracking@iop.org

Additional copies of this publication are available from:

Curran Associates, Inc.
57 Morehouse Lane
Red Hook, NY 12571 USA
Phone: 845-758-0400
Fax: 845-758-2633
Email: curran@proceedings.com
Web: www.proceedings.com

PREFACE

It is our great pleasure to introduce the proceedings of the 1st International Conference on Advanced Research on Engineering Sciences and Technology (ICAREST 2021) that has been organized by Al-Mustaqbal University College, Iraq, and scientifically sponsored by eSystems Engineering Society, UK, from the 6th to the 8th of April, 2021.

The core aim of ICAREST 2021 has always been to bring together early-career researchers, scientists, academics, engineers, and postgraduate students to exchange and share their experiences and research results on all aspects of engineering sciences and technology. The professional keynote speakers and researchers have presented various perspectives on research.

The scientific theme for ICAREST 2021 is "Engineering Researches are Essential Pillar of Scientific and Industrial Development". The aim of this conference is to capture the substantial role that advanced engineering technology plays in addressing many challenges such as the world's demands for energy, environmental protection, communication networks developments, manufacturing processes developments, etc. with both a focus on the fundamentals (foundations) and future directions (vistas).

ICAREST 2021 covers different tracks, including Civil Engineering, Chemical Engineering, Materials Engineering, Mechanical Engineering and Power Engineering, Computer Engineering and Information Technology, Communication Systems and Networking, as well as Biomedical Engineering. After a peer-review process, around 28 scientific papers are finally accepted out of 218 total submissions. We are sure that the proceedings will serve as an important research source of references and knowledge, which will lead to demonstrate novel ideas and impressive efforts in engineering sciences and technology.

We wish to express our sincere gratitude to our partners, authors, reviewers, and committees of ICAREST 2021 for their tremendous contribution to make this international event hosted successfully in this challenging time.

Finally, we hope to meet you all in person at our next conference.

Warmest regards,

Asst. Prof. Hasan Shakir Majdi (Dean of Al-Mustaqbal University College, Iraq)

IOP Conf. Series: Materials Science and Engineering 1184 (2021) 011001 doi:10.1088/1757-899X/1184/1/011001

COMMITTEES LIST

Honorary General Chair

Prof. Kamaruzzaman Sopian National University of Malaysia, Malaysia.

Local General Chair

Asst. Prof. Hasan Sh. Majdi, Al-Mustaqbal University College, Iraq.

Honorary General Co-Chair

Prof. Muthanna Al-Dahhan, Missouri University of Science and Technology, USA.

General Co-Chair

Asst. Prof. Azher M. Abed, Al-Mustaqbal University College, Iraq.

Scientific Committee Chair

Prof. Mohammed H. Al-Maamori, University of Babylon, Iraq.

Steering Committee Chair

Dr. Hamza Mohammed Ridha Al-Khafaji, Al-Mustaqbal University College, Iraq.

Publicity Chair

Dr. Ali Majdi, Al-Mustaqbal University College, Iraq.

Scientific Committee Members

- Asst. Prof. Abdollah Bahador, Japan
- Prof. Abdulmajeed A. Mohamad, Canada
- Dr. Ali Ibrahim Moslem, Hungary
- Asst. Prof. Ammar zeidan Ghailan, Turkey
- Dr. Abbas J. Sultan, Iraq
- Asst. Prof. Abdelsalam Efhaima, UAE
- Prof. Alaa Al-Hamami, Jordan
- Asst. Prof. Ali Naji Attiyah, Iraq
- Assoc. Prof. Ammar Muthanna, Russia
- Assoc. Prof. Anastasia A. Minina, Russia
- Prof. Basim Hasan Obaid, Iraq
- Prof. Ehsan Noroozinejad Farsangi, Iran
- Asst. Prof. Essam Zuhair Fadhil, Iraq
- Asst. Prof. Ghassan Fadhil M. Hussain Smaisim, Iraq
- Prof. G. M. Sadiqul Islam, Bangladesh
- Dr. Hassnen Jafer, Iraq
- Dr. Hayder Kamil Shanbara , Iraq
- Prof. Hidayah Sulaiman, Malaysia
- Dr. Humayun Shariff, USA
- Asst. Prof. Ibrahim Ahmed Siad, Egypt
- Assoc. Prof. Ilinca Nastase, Romania
- Prof. Imad Fakhri Al Shaikhli, Malaysia
- Asst. Prof. Khaled A. AL Farhany, Iraq
- Prof. Saad Saffah Hreshee, Iraq
- Dr. Lateef Assi , Iraq
- Asst. Prof. Mahmoud Taha, Egypt
- Prof. Mohammad Alghoul, KSA
- Dr. Mohd Fauzi Bin Mamat, Malaysia
- Prof. Mohammed ASSI Ahmed AL-dujaili, Iraq

- Asst. Prof. Mohammed K. Al Mesfer, KSA
- Assoc. Prof. Mohsin Abbas Aswad, Iraq
- Assoc. Prof. Muhammad Ramlee Kamarudin, Malaysia
- Prof. Nada Sadoon, Iraq
- Prof. Nadhir Al-Ansari, Sweden
- Asst. Prof. Nasser Ibrahim Zouli, KSA
- Asst. Prof. Neven Ali, USA
- Prof. Qusai Fadil Al-Salihi, Iraq
- Asst. Prof. Qusay Rasheed Abed Al-amir, Iraq
- Asst. Prof. Rahman almusafir, USA
- Assoc. Prof. Rami Hikmat Al-Hadeethi, UK
- Prof. Syed Alwee Aljunid, Malaysia
- Prof. Ahmed T. Sadiq, Iraq
- Dr. Safaa Najah Saud, Malaysia
- Assoc. Prof. Shamsuddin Shahid, Malaysia
- Assoc. Prof. Tuty Asma Abu Bakar, Malaysia
- Dr. Vafa Soltangharaei, USA
- Dr. Yue Huang, UK
- Dr. Zaher Mundher Yaseen, Vietnam
- Asst. Prof. Sarmad Talib Najim Al-Ani, Iraq
- Asst. Prof. Akeel Abdullah Mohammed, Iraq
- Asst. Prof. Khalid Abdul Wahhab Mustafa, Iraq
- Asst. Prof. Haitham Alaa Hussein Alshami, Iraq
- Asst. Prof. Maher Yahya Salloom, Iraq
- Asst. Prof. Ameer H. Morad, Iraq
- Asst. Prof. Dhirgham Kamal Naji, Iraq
- Dr. Ali H. Al-Timemy, Iraq
- Dr. Nebras H. Ghaeb, Iraq
- Dr. Mahmoud Ibraheem Al-Kadi, Iraq
- Dr. Noor Kamal Mohsin Al-Qazzaz, Iraq
- Asst. Prof. Dr. Jamal Manie Ali

Steering Committee Members

- Asst. Prof. Ali Najah Al Shamani, Iraq
- Mr. Amir Najah Saud, Iraq
- Dr. Alaa Mohammed Hussein Wais, Iraq
- Dr. Ali Shubbar, UK
- Mr. Ammar Abdulkadhim, Iraq
- Asst. Prof. Essam Kamel, Iraq
- Dr. Khalid Omran Ali, Iraq
- Dr. Laith Habeeb, Iraq
- Dr. Laith S. Sabri, Iraq
- Mr. Luay Hashem, Iraq
- Asst. Prof. Malik Mustafa, Iraq
- Dr. Maryam Hameed Al-Mamoori, Iraq
- Asst. Prof. Wael Al-Nidawi, Iraq
- Mr. Tameem Mohammed Hashim, Iraq
- Dr. Ameen M. Al-Juboori, Iraq
- Mr. Musaddak Maher Abdulzahra, Iraq
- Ms. Mayadah Waheed Falah, Iraq
- Dr. Mohammed Zuhear Almulali, Iraq

IOP Conf. Series: Materials Science and Engineering 1184 (2021) 011002 doi:10.1088/1757-899X/1184/1/011002

Peer review declaration

All papers published in this volume of *IOP Conference Series: Materials Science and Engineering* have been peer-reviewed through processes administered by the Editors. Reviews were conducted by expert referees to the professional and scientific standards expected of a proceedings journal published by IOP Publishing.

- **Type of peer review:** Double-blind, the authors do not know who the reviewers are.

- **Conference submission management system:** EasyChair

  Submission link: https://easychair.org/conferences/?conf=icarest2021

- **Number of submissions received:** 218

- **Number of submissions sent for review:** 153

- **Number of submissions accepted:** 28

- **Acceptance Rate (Number of Submissions Accepted / Number of Submissions Received × 100):** 12.84 %

- **Average number of reviews per paper:** 2 to 3

- **Total number of reviewers involved:** 102

- **Any additional info on the review process:**

During the review process, all papers were checked to ensure that:

1- The paper must be relevant to conference topics;
2- The paper should be written in accordance with the standard template of the IOP;
3- Papers are written in English with acceptable language;
4- Papers contain good quality graphic images;
5- The maximum percentage of plagiarism permitted is 20% (using Turnitin);
6- All papers that passed the similarity index and have already followed the IOP template will be sent to the reviewers;
7- Reviewers reviewed the papers by checking the scientific value and give comments to the papers;
8- The scientific committee decided whether the papers: accepted, accepted with revision, or rejected;
9- The authors revised their papers and sent back their revisions for the final decision.

- **Contact person for queries:**

  Dr Khalid Hashim, email: eng.khalid.saffa@uobabylon.edu.iq

Dr Ali Shakir, email: alimajd@mustaqbal-colledge.edu.iq

Economic Feasibility Study of a Hybrid Power Station Between Solar Panels and Wind Turbine with The National Grid in Al-Hayy City in the Central of Iraq

Ahmed Z. Abass[1,*], D.A Pavlyuchenko[1], Haider Th. ALRikabi[2], Faisal T. Abed[2], Julian Gaidukov[1]

[1] Novosibirsk State Technical University, Novosibirsk, Russia
[2] Electrical Engineering Department, College of Engineering, Wasit University, Wasit, Iraq

* Corresponding author: theking.amir@mail.ru

Abstract. In view of the increasing demand for electric energy, parallel to global warming and increasing temperatures in the world, efforts must be made to find new and environmentally friendly ways to produce electric energy and eliminate the burning of fossil fuels. In this paper, we propose to approve hybrid plants for the production of electric power in Iraq due to the availability of the privileged location and the amount of clean energies that can be invested. We have chosen a small area located in the south of Iraq and suggested the establishment of a hybrid plant between solar energy, wind, and the national grid, and the results were very impressive, where at a very low cost it is possible to cover the district's need of electricity and export the rest to the national grid at a price (0.015 kwh). We chose this type of station, wherein in the daytime, it is possible to take advantage of solar energy, as well as wind energy, if available. And wind energy can be used at any time when there is wind at a speed higher than 3 m/s, where energy can be generated at night hours and the absence of sunlight for any reason. This study was carried out with the help of the NASA space website and the Homer program to study renewable energies.

Keywords: Renewable energy, hybrid plant, solar energy, wind energy.

1. Introduction

"Al- hayy district" is an Iraqi town and a district center in Wasit Governorate. The city's population is estimated at 171,000, while the district is approximately 280,000, along with the surrounding villages and sub-districts. The city has two sub-districts (Al-Bashaer and Al-Mawfakia)[1]. These two areas are distinguished by their good geographical location and the abundance of easy and flatlands. In addition, the city of Al- hay suffers, like the rest of the other cities of Iraq, from poor and shortage of electric power. In this paper, we will deal with the sub-district of Al-Bashayer as an example that can be applied to the rest of the neighboring cities, and how the city can be made self-sufficient in generating electrical energy through the exploitation of available clean energy resources, as well as the possibility of the city being an energy source for neighboring sub- districts and villages in some times. The problem of the shortage in the provision of electrical energy in Iraq, beginning in 1991 after the US-Iraq war, when the American forces bombed most of the Iraqi installations, including the electrical infrastructure such as the generation, transmission, and distribution station[2-8]. Look at the figure below:

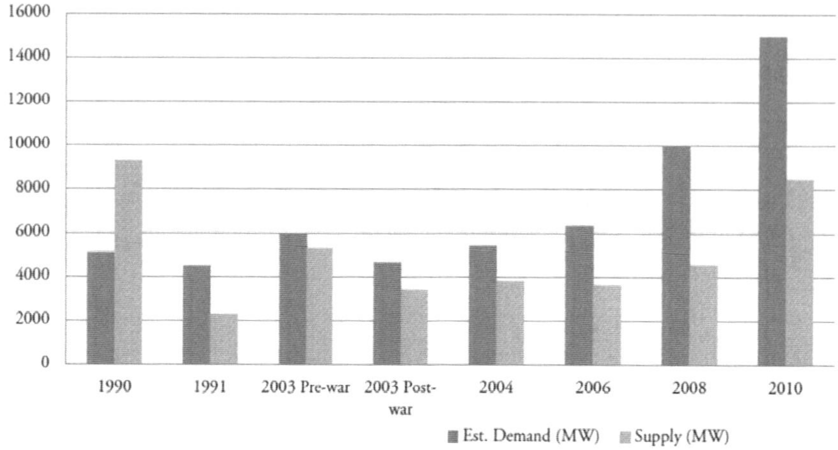

Figure 1. Electricity supply/demand (1991 – 2010)

From this figure clear that Iraq has a big gap between electric supply and demand and this gap increasing with time Because of the increasing consumer demand for the purchase of electrical appliances as well as the development of the country in the field of industry. The Iraqi MoE has specified the shortage in the production of electricity more than 6000 MW from May to the September[9-12].

2. Weather in Al-Hayy

Al-summers hayy's are long, hot, arid, windy, and clear, while the winters are cool, dry, and mostly clear. The temperature typically ranges from 7 to 46 degrees Celsius throughout the year, with temperatures rarely falling below 3 degrees Celsius or rising above 49 degrees Celsius. The graph below depicts a summary of climate change[13-16].

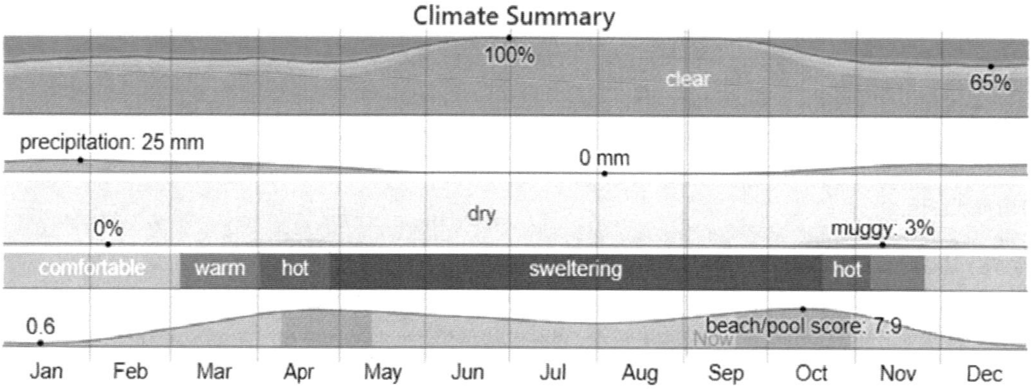

Figure 2. Climate in Al-hayy

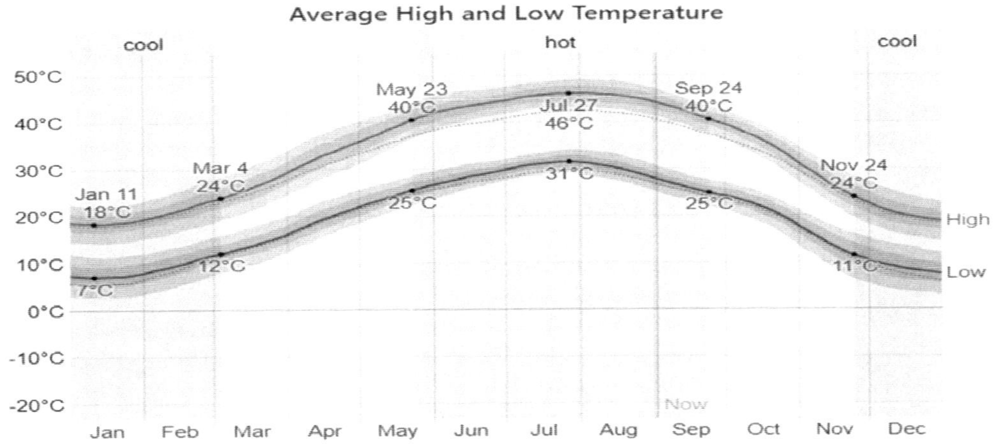

Figure 3. average of temperature in Al-hayy

The average percentage of cloudy sky in Al-hayy varies significantly throughout the year due to seasonal variations. In Al-hayy, the clearer season begins around May 22nd and lasts for 4.7 months, ending around October 13th. The sky is completely obvious, mostly concise, or cloudy weather 100% of the time on June 30th, the clearest day of the year, and cloudy or mostly stormy 0% of the time. The cloudier season begins around October 13th and lasts for 7.3 months, ending around May 22nd. As a result, the clear weather encourages people to invest in solar panels; the graph below depicts the sky over a year.

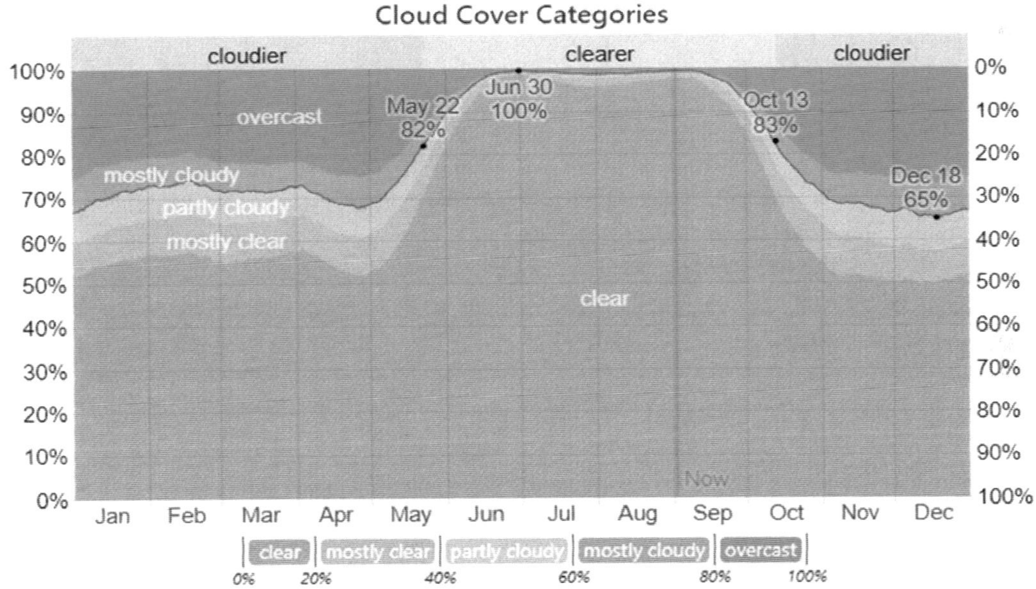

Figure 4. Clearness sky in Al-hayy

2.1. Solar Energy and Hours of Daylight in Al-Hayy

This section will discuss the total daily infraction solar energy reaching the ground over a large area in Al-hayy, taking into account seasonal variations in the length of the day, the elevation of the Sun above the horizon, as well as absorption by clouds and another atmospheric component, as mentioned earlier. Shortwave radiation includes visible light and ultraviolet radiation. The mean daily incident solar energy experiences extreme seasonal variation over the year. The brighter time of the year continues for 3.4 months, from May 14th to August 25th, with average daily incident energy above 7.3 kWh/m^2. The brightest day of the year is June 19th, with an average of 8.3 kWh/m^2 The darker part of the year continues for 3.2 months, from the first of November to February 8th, with a mean daily incident energy of 4.1 kWh/m^2[17].

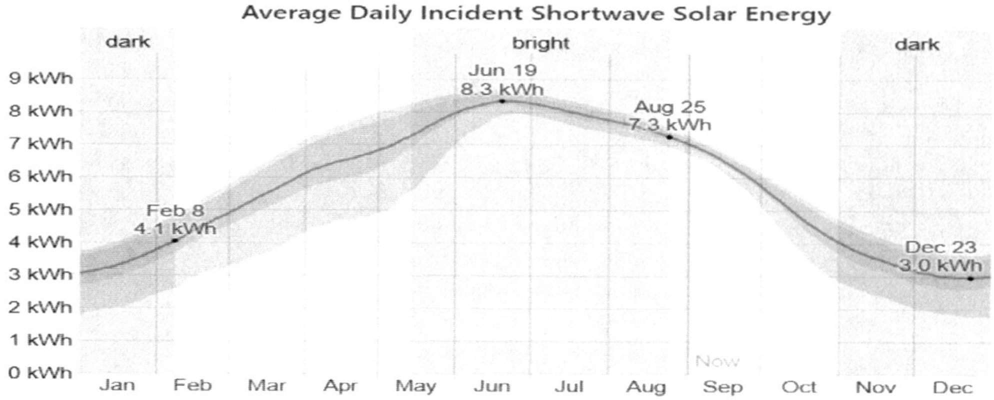

Figure 5. Average solar energy in Al-hayy

Table1. Average solar energy in Al-hayy [7]

LAT	LON	PARAMETER	JAN	FEB	MAR	APR	MAY	JUN	JUL	AUG	SEP	OCT	NOV	DEC	ANN
32.25	46.25	DNR	5.16	5.8	5.51	5.49	6.75	8.66	8.1	7.64	6.9	4.8	4.14	4.46	6.12
32.25	46.25	DNR_MAX	6.27	7.85	7.11	6.72	8.37	10.74	9.6	8.65	7.83	5.87	4.98	5.17	7.43
32.25	46.25	DNR_MIN	3.59	4.2	4.45	4.04	5.64	6.48	6.12	6.61	5.32	3.87	3.34	3.2	4.74

The length of the day in Al-hayy varies greatly throughout the year. The tightest day in 2020 is December 21st, which has ten hours and two seconds of daylight, and the lengthiest day is June 21st, which has fourteen hours and sixteen minutes of daylight. The diagram below depicts the hours of daylight:

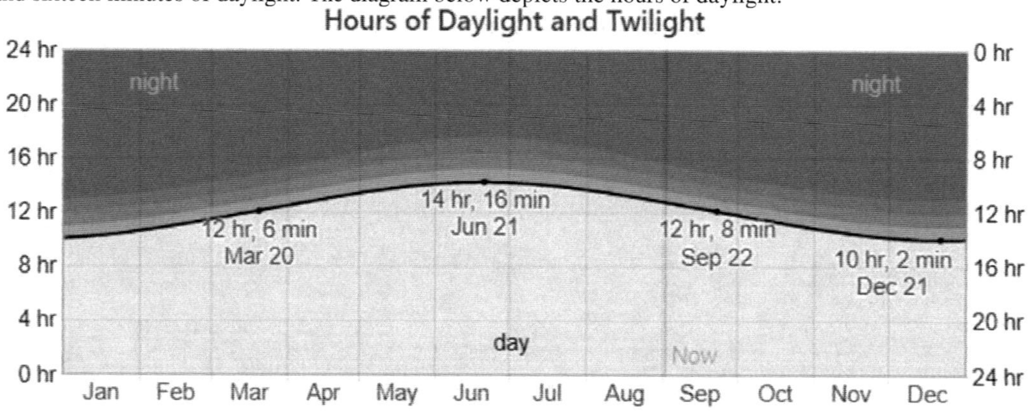

Figure 6. the average of hours daylight in Al-hayy

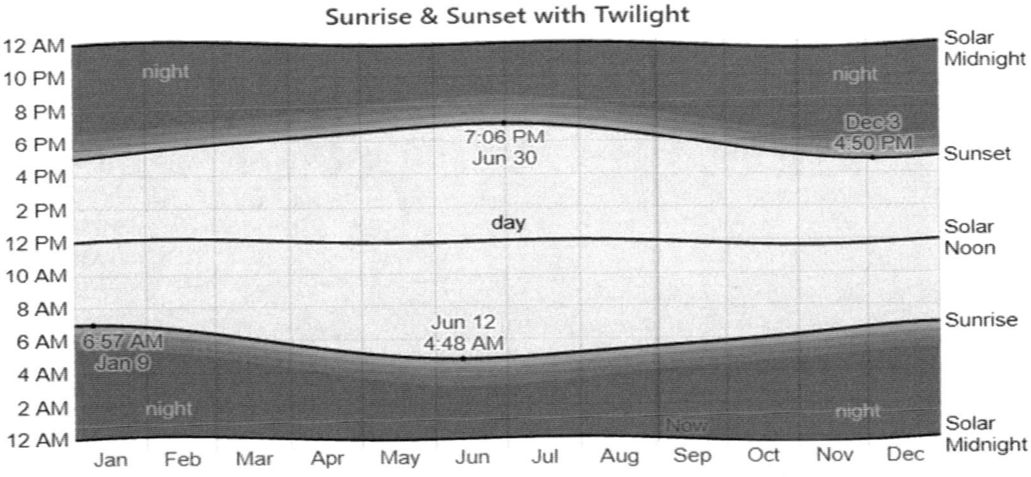

Figure 7. The solar day over the year in Al-hayy (2020)

2.2. Wind Energy in Al-Hayy

This part explains the wide-area hourly rate speed of wind at ten meters above the land. The wind experienced at any specific location is based on local topography and other elements, and instantaneous wind (speed &direction) varies more widely than hourly rates. The rate hourly wind speed in the Al hayy having significant seasonal variation over the year. The most windier time of the year continues for 3.5 months, from May 26th to September 8th, with mean wind speeds of more than 5 m/s[18]. The other time of year lasts for more than 8 months, from September 8th to May 26th with an average hourly wind speed between (3.5m/s to 4.5m/s). The figure below shows the average wind speed in Al-hayy:

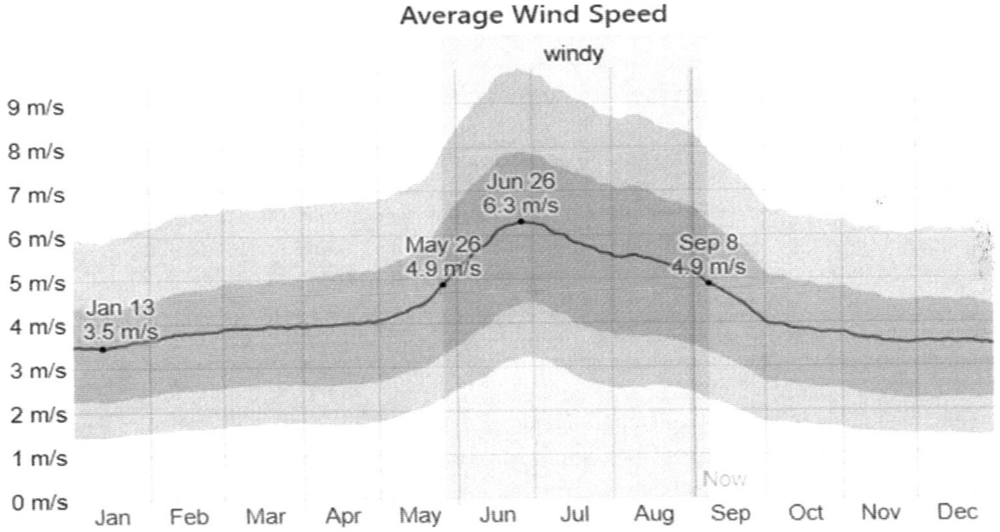

Figure 8. average wind speed in Al-hayy

2.3. Case Study and Calculation Method

In this paper, we will deal with one of the small sub-districts of the Al-hayy city, as mentioned earlier, and it is called Al-Bashaer. The load in this sub-district is distributed between the center and the surrounding villages, and the load ranges from 800 kW to 1.3 MW, the peak load in summer. By using some programs and websites specialized in monitoring and following up space and weather conditions, such as the NASA space organization, we surveyed the city about the quantities of clean energy sources in the city and how they are used, and their economic viability. And by using the famous program HOMER [19-21], which is specialized in calculating all types of renewable energies. As well as the use of important mathematical equations and we will mention here the most important of these equations, HOMER uses the following equation to calculate the output of the PV array:

3. Simulation and Results

By using the Homer program, we were able to obtain wonderful results that are close to reality, and this study helps us to get rid of the dire reality of electrical energy and to identify the volume of clean energy that is not being used properly. The proposed station is a hybrid between solar energy, wind, and the national grid. The solar station consists of a solar panel station with a capacity of 540 kW (Schneider ConextCoreXC 540kW with Generic PV), distributed over an area of about 100 acres. The wind station is of the type (Vergnet GEV MP-C [275kW])[12, 22, 23]

- 2-blade downwind rotor, two-speed generator
- Cut-in wind speed is 3.5 m/s
- Cut out wind speed is 25 m/s
- Output Voltage & Frequency (3-phase are 400 V - 50 Hz respectively
- Hub height is 55/60 m
- Rotor diameter is 32 m
- Rotation speed (50 & 60 Hz) is from 31 to 46 rpm.

Looking at figure 9 gives us a clear vision of the power plant and its components and connecting it to the national network [8].

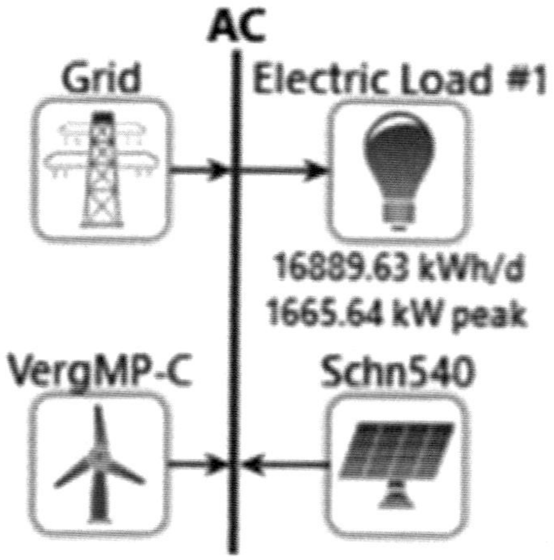

Figure 9 Single diagram for the system

After entering all the data about the city Al-hayy and sub-district Al-Bashayr from the geographical location and daily loads, we get the following results electrical summary

Table 2. Fabrication Summary

Component	Production (kWh/yr)	Percent
Schneider ConextCoreXC 540kW with Generic PV	32,888,049	20.07
Vergnet GEV MP-C [275kW]	65,965,943	78.4
Grid Purchases	1,417,801	1.55
Total	91,271,793	100

Table 3. Consumption Summary

Component	Consumption (kWh/yr)	Percent
AC Primary Load	6,164,453	6.75
DC Primary Load	0	0

Grid Sales	85,107,340	93.2
Total	91,271,793	100

Figure 10 shows the production of electrical energy, where it appears that the highest percentage is for wind energy, as it constitutes about 70% of the total product, and then solar energy production is in second place by about 20%, and the rest is provided by the national grid.

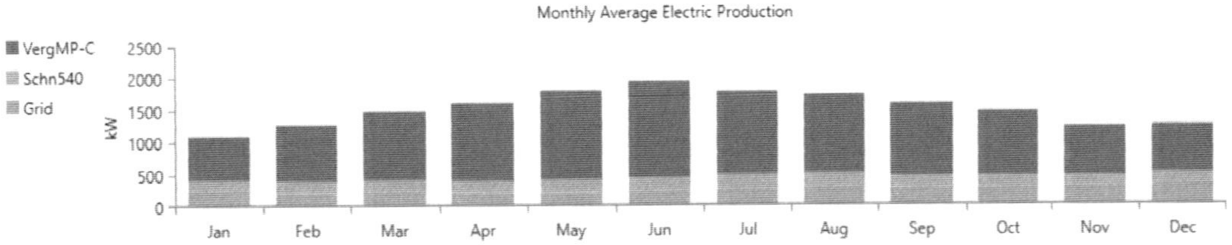

Figure 10. Average electric production

Tables4 and 5 a summary of the production of the solar plant (Schneider ConextCoreXC 540kW with Generic PV), as well as figure 11.

Table 4. Generic PV Electrical summary

Quantity	Value	Units
Minimum Productivity	0	kW
Optimum Productivity	540	kW
PV Penetration	30.6	%
Operation Hours	4,387	hrs/yr
Level of Cost	0.0371	$/kWh

Table 5. Generic PV Statistics

Quantity	Value	Units
Rated Capacity	2,111	kW
Mean Output	216	kW
Mean Output kwh/day	5,173	kWh/d
Capacity Factor	10.2	%
Total Production	32,888,049	kWh/yr

Figure 11 shows the production of electrical energy with the help of solar cells, which is limited during daylight hours only. It is also clear through the colors that it is produced during the summer months more than the rest of the year, where the color is more clear and deep during the summer than the rest of the year

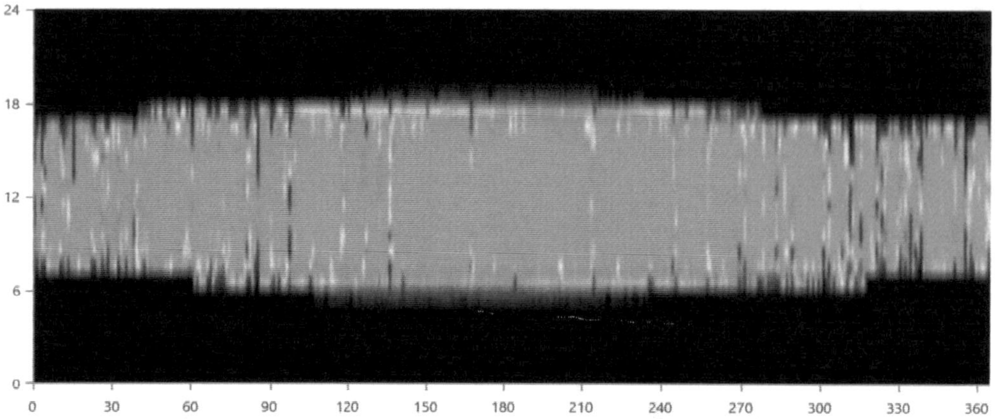

Figure 11. solar plant PV Output (kW)

In tables 6 and 7 a summary of the production of the wind turbine (Vergnet GEV MP-C [275kW]), as well as figure 12.

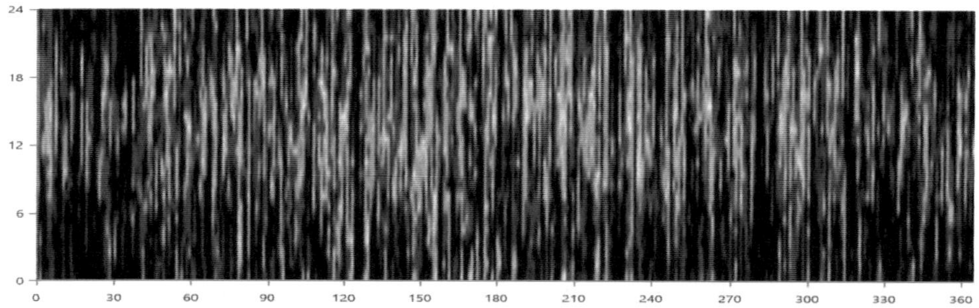

Figure 12. Wind turbine output (kW)

Table 6. Net energy purchased and sold in grid

Month	Energy Purchased (kWh)	Energy Sold (kWh)	Net Energy Purchased (kWh)	Peak Demand (kW)	Energy Charge
January	174,827	373,805	-198,978	889	$43,333
February	135,823	453,015	-317,192	1,040	$33,503
March	152,419	616,709	-464,291	1,082	$37,488
April	132,720	704,178	-571,457	1,024	$32,476
May	139,972	829,675	-689,704	1,110	$34,163
June	146,608	846,050	-699,442	1,200	$35,806
July	176,338	710,671	-534,334	1,192	$43,374
August	199,608	679,603	-479,995	1,200	$49,222
September	177,284	612,059	-434,775	1,119	$43,709
October	197,050	547,864	-350,814	1,200	$48,715
November	206,641	361,470	-154,828	1,064	$51,299
December	241,501	377,783	-136,282	1,186	$59,998
Annual	2,080,791	7,112,883	-5,032,092	1,200	$513,085

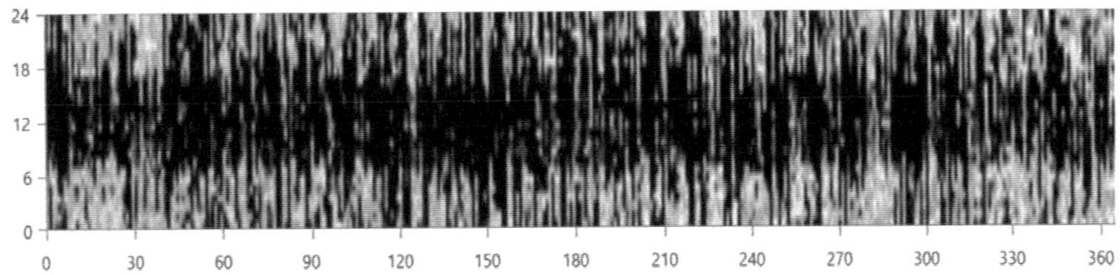

Figure 13. Energy bought from the grid (kW)

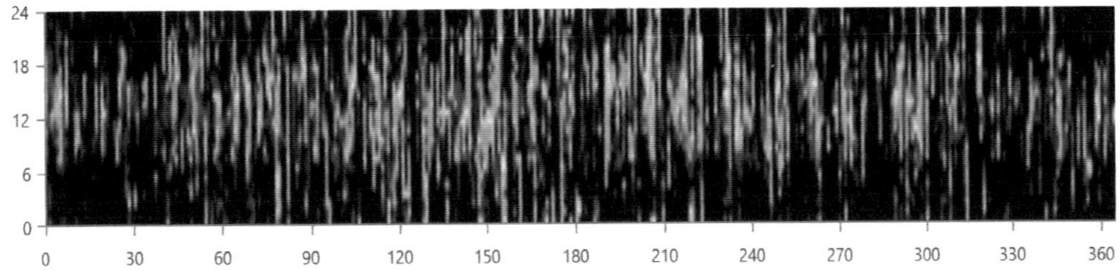

Figure 14. Energy vended to the grid (kW)

4. Conclusion

Given the privileged location of Iraq, which is rich in clean energy resources, and because of the positive results obtained from this study, clean energies must be invested in Iraq. This idea should be spread throughout Iraq, where clean energy can be produced at a very cheap cost (0.015$ kwh), get rid of burning fossil fuels, and create new jobs in the region that suffers from high unemployment. When looking at table 8, we see the amount of energy exported to the national grid, and throughout the year the energy exported is more than imported, and this can be seen from figures 12 and 13. It appears in table 4 that the number of working hours of the solar plant is about 4386 hours, which is a very large number, and this can be seen from figure 12, where most of the energy exported to the grid is during daylight hours. Table 6, shows us the number of working hours of the wind turbine, which is about 6414 hours, which is also a very large number, and this can be seen in figure 12, showing us the amount of energy produced during the year. This shows us the economic feasibility of this type of station

References

[1] Issa M J Study of Boron, Phosphate, Nitrate and H2S Gas Pollution in the Water of Tigris River From Al Suwayra to Al Hayy City, Wasit, Iraq

[2] Abass A Z and Pavlyuchenko D 2019 Southern Iraq gas station conversation to integrated solar combined cycle. In: *E3S Web of Conferences*: EDP Sciences) p 05008

[3] Zubaidi S L, Abdulkareem I H, Hashim K S, Al-Bugharbee H, Ridha H M, Gharghan S K, Al-Qaim F F, Muradov M, Kot P and Al-Khaddar R J W 2020 Hybridised Artificial Neural Network Model with Slime Mould Algorithm: A Novel Methodology for Prediction of Urban Stochastic Water Demand **12** 2692

[4] Hashim K S, Ewadh H M, Muhsin A A, Zubaidi S L, Kot P, Muradov M, Aljefery M, Al-Khaddar R J W S and Technology 2021 Phosphate removal from water using bottom ash: Adsorption performance, coexisting anions and modelling studies **83** 77-89

[5] Alrikabi H, Alaidi A H and Nasser K 2020 The Application of Wireless Communication in IOT for Saving Electrical Energy *International Journal of Interactive Mobile Technologies* **14** 152-60

[6] Tuama H, Abbas H, Alseelawi N S and ALRikabi H T S 2020 Bordering a set of energy criteria for the contributing in the transition level to sustainable energy in electrical Iraqi Projects *Periodicals of Engineering and Natural Sciences* **8** 516-25
[7] Roa'a M A a, Aljazaery I A, Al_Dulaimi S K, Alrikabi H T S and Informatics 2021 Generation of High Dynamic Range for Enhancing the Panorama Environment *Bulletin of Electrical Engineering* **10**
[8] Hussein A S, Khairy R S, Najeeb S M M and ALRikabi H T J I J o I M T 2021 Credit Card Fraud Detection Using Fuzzy Rough Nearest Neighbor and Sequential Minimal Optimization with Logistic Regression **15**
[9] Abass A Z, Pavlyuchenko D J I J o E and Engineering C 2019 The exploitation of western and southern deserts in Iraq for the production of solar energy **9** 4617
[10] Tuama H, Abbas H, Alseelawi N S and AlRikabi H T H S 2020 Bordering a set of energy criteria for the contributing in the transition level to sustainable energy in electrical Iraqi projects *Periodicals of Engineering and Natural Sciences* **8** 516-25
[11] Zubaidi S L, Al-Bugharbee H, Muhsin Y R, Hashim K and Alkhaddar R 2020 Forecasting of monthly stochastic signal of urban water demand: Baghdad as a case study. In: *IOP Conference Series: Materials Science and Engineering*: IOP Publishing) p 012018
[12] Khazaal H F, Alrikabi H T S, Abed F T and Kadhm S I 2019 Water desalination and purification using desalination units powered by solar panels *Periodicals of Engineering and Natural Sciences* **7** 1373-82
[13] Abass A, Pavlyuchenko D and Hussain Z S 2020 Methods Comparison for Optimal Capacitor Placement in Distribution System. In: *2020 International Multi-Conference on Industrial Engineering and Modern Technologies (FarEastCon)*: IEEE) pp 1-6
[14] Abass A Z, Pavlyuchenko D, Hussain Z S and Control 2021 Survey about impact voltage instability and transient stability for a power system with an integrated solar combined cycle plant in Iraq by using ETAP *Journal of Robotics* **2** 134-9
[15] Ethaib S, Omar R, Kamal S M M, Awang Biak D R and Zubaidi S L J P 2020 Microwave-assisted pyrolysis of biomass waste: a mini review **8** 1190
[16] Mohammed B K, Mortatha M B, Abdalrada A S, ALRikabi H T S J P o E and Sciences N 2021 A comprehensive system for detection of flammable and toxic gases using IoT **9** 702-11
[17] Abed F T, ALRikabi H T S and Ibrahim I A 2020 Efficient Energy of Smart Grid Education Models for Modern Electric Power System Engineering in Iraq. In: *IOP Conference Series: Materials Science and Engineering*: IOP Publishing) p 012049
[18] Sayago S, Ovando G, Almorox J and Bocco M J I J o R S 2020 Daily solar radiation from NASA-POWER product: assessing its accuracy considering atmospheric transparency **41** 897-910
[19] Kumar S, Upadhyaya P and Kumar A 2019 Performance Analysis of Solar Energy Harnessing System Using Homer Energy Software and PV Syst Software. In: *2019 2nd International Conference on Power Energy, Environment and Intelligent Control (PEEIC)*: IEEE) pp 156-9
[20] Naseer Ali Hussien A A D A-m, Haider Th.Salim Alrikabi, Faisal Theyab Abed 2021 Monitoring the Consumption of Electrical Energy Based on the Internet of Things Applications *International Journal of Interactive Mobile Technologies (iJIM)* **15**
[21] Al-dabag M, ALRikabi H S and Al-Nima R 2021 Anticipating Atrial Fibrillation Signal Using Efficient Algorithm *International Journal of Online and Biomedical Engineering (iJOE)* **17** 106-20
[22] Romanov A, Hatckevich A and Vyrmaskin D J П н м 2020 Anwendungsentwicklung Umgebung von Schneider Electric für industrielle Steuerungen **1** 22-5
[23] Hala A. Naman N A H, Mohand Lokman Al-dabag, Haider Th.Salim Alrikabi 2021 Encryption System for Hiding Information Based on Internet of Things *International Journal of Interactive Mobile Technologies (iJIM)* **15**

Internet of Things (IoT) application in the assessment of learning process

Nabaa A. Jasim[1,*], Haider Th. Salim ALRikabi[1], Mansour S. Farhan[1]

[1] Electrical Engineering Department, College of Engineering, Wasit University, Iraq

* Corresponding author: hdhiyab@uowasit.edu.iq

Abstract. The spread of Coronavirus has changed the common life of students over the world. As a result of following international guidelines, all educational institutions are required to work remotely to reduce the contagion of the spread of Coronavirus. Web-based technologies and the Internet have become popular platforms for developing and introducing distance learning programs in the online classroom. Therefore, an application designed specifically to support e-Learning effectiveness named E-Learning Motivation Systems (E-LMS) changes in the teaching-learning process, monitoring and tracking of student performance. this article aims to identify and propose a low-cost, effective and flexible platform, This platform is designed to monitoring preliminary data on student access to, and participation in, virtual classroom environments and activities. In an initial case study conducted at the electrical engineering department at Wasit University. According to the evaluation results, E-LMS Motivation is an effective support tool for educators in pedagogical monitoring and student performance analysis, significantly contributing to improved student retention and promotion rates

Keywords: Raspberry Pi, e-learning, Internet of things (IoT), an education platform

1. Introduction

Education is one of the most perceptible human activities affected by the Internet[1-3]. Distance education (DE) is being presented as a solution to meet the new educational demands arising from globalization, which is not only an economic phenomenon but also a transformation of space and time. The availability of technological resources alleviates the difficulties associated with the physical distance between students and teachers by allowing the creation of a virtual environment that promotes collaborative learning. As a result of the wide access to the Internet, online education enables professionals and students to study from a distance and keep contacting with the newest technologies and administrative changes[4-6]. E-learning has developed as one of the best moving trends in advanced learning, especially at present after the emergence of the Corona virus[7, 8]. Where now electronic teaching is taking place, in addition to providing thousands of training, technical and administrative courses, as well as degree and certificate programs by universities, professional development centers, and industrial training facilities all over the world[9, 10]. Smart learning is one of the main elements in the implementation of smart cities, and it is activated using virtual learning and digitalization[11-15]. This classification has led to many investigations by researchers. The Internet and Web-based technologies became a popular platform for developing and delivering distance learning (DL) programs in an online class setting in 2000. The E-LMS Motivation is viewed in this context as a method for monitoring students' success in the E-LMS Moodle[16, 17]. The aim is to help teachers, staff, and advisors with the role of advising students. Keeping track of distance learning students' progress. For this reason, Validation of the E-LMS Motivation program and student data.

2. Background about Internet of Things (IoT)

It mentions to networking technical innovations that allow real-world objects to join with one another by the internet. The devices that are linked are referred to as 'things'[13, 18-23]. The ability to send and receive information through the interconnection of various things via the internet has numerous applications in almost every field, including healthcare, business, transportation, agriculture, management, and education [24, 25]. These technologies provide a variety of e-learning technologies that can change the future of education systems[26]. The (IoT) in general, as well as some of its applications, will be discussed in this paper, with an emphasis on E-learning. Therefore, researchers **Luvai Motiwalla and Steven Tello** created a Web-based course model and analysed the results of an exploratory study to see how effective it was at increasing student satisfaction. The outcomes of their surveys presented that students were mostly gratified with the virtual education situation and demonstrated their appreciation for the suppleness of being ability to access the course materials anytime and anywhere, and they were satisfied with the quality and content of the course[27]. Both researchers, **Keenan A. Pituch** and **Yao-Kuei Lee** (in 2006) suggested and tested alternative models aimed at explaining the student's intention to use the e-learning organization when using the system as a auxiliary educational tool within the traditional classroom or an independent method of distance education. The determinants of the established technology acceptance model, as well as the system and participant characteristics mentioned in the research literature, were incorporated into these models. Data were collected by 259 university students, to calculate the stage of presentation and use of the e-learning system. Structural equation modeling provided better support for a model that assumed stronger effects of system properties on the use of an e-learning system, in addition to discussing the implications for both researchers and practitioners[28]. In 2009, many universities were applying e-learning for various reasons. Based primarily on the technology acceptance model, Sung Youl Park proposed a comprehensive analytical foundation for university students' acceptance and intention to use e-learning (TAM). Huge enquiry efforts were devoted to developing e-learning systems, and several e-learning systems were suggested and used in practice, but the completion rate of e-learning was low in these systems at that time. Therefore, the author **Masafumi Yamada et al** in 2016 designed and tested Internet-based e-learning to improve the performance of the system by the researcher. Their proposed system can motivate the learner by using the smart implemented box, microwave sensor, and decision tree[29]. Some researcher indicated that the current semester's teaching absences exploration of students' information, which is incapable of evaluating students' mastery of knowledge points and locating students' learning paths in real-time, as well as effectively recommending the necessary achieve material for students. So, they proposed a personal familiarity point endorsement system based on the course data chart to solve these problems[30, 31]. Using their proposed system, they can first create evaluation equations to judge students' mastery of knowledge points, by using accuracy, response time, and answer types in the answer records. Secondly, they can extract the knowledge points in answer records and map them into the course knowledge graph. Finally, they can recommend knowledge points assigned to students using the semantic ordered association and the knowledge point sequence in the knowledge graph [18]. In this article, the focus was on the possibility of motivating students to pursue their studies by proposing a platform to monitor their school bikes and their daily attendance by the student himself and his parent. Where the teachers register this data on the platform on Raspberry Pi to be displayed to students and parents.

3. Proposed work

This paper proposes an easy, simple, and reliable online method for monitoring the student's study status by the student himself and his parent. The E-LMS Motivation tool was created as a Moodle E-LMS plugin, and it looks like a block with options inserted in a virtual space. This is done by designing and implementing an educational platform in which the student's academic grades and daily attendance are written by the responsible teacher. This platform is built on the server is the Raspberry Pi shown in figure (1)

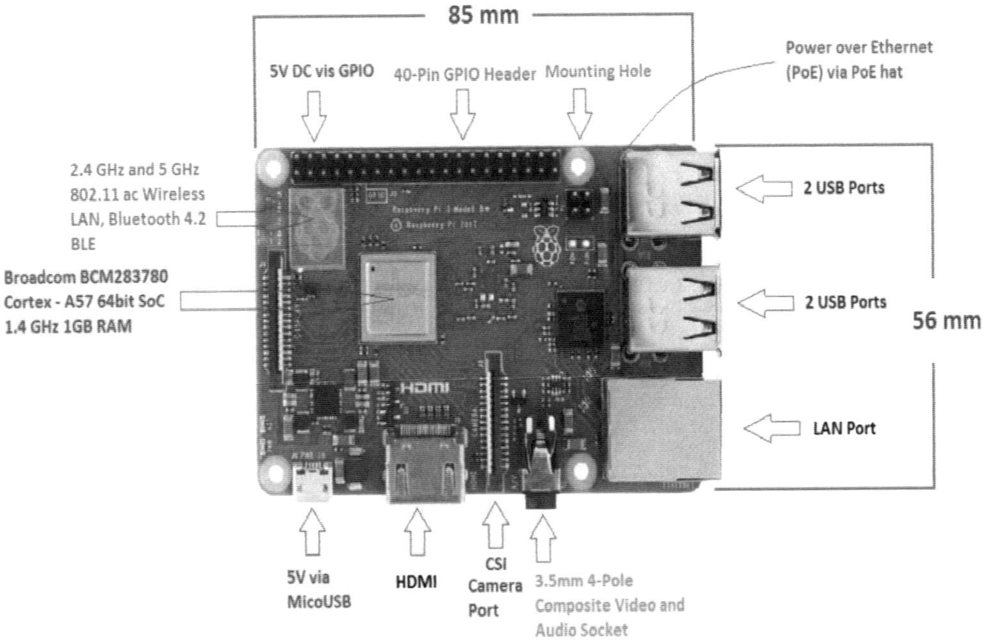

Figure 1. The Raspberry Pi 3.

The **Raspberry Pi :-** It's a tiny computer board with a CPU, GPU, USB ports, Wi-Fi, Bluetooth, and a set of GPIO (general purpose input/output) pins for controlling electronic components for physical computing and experimenting with the Internet of Things (IoT). The server (Raspberry Pi) is built using the things board application. Thingsboard is an open-source Internet of things (IoT) platform that enables rapid development, management, and expansion of IoT projects. Figure (2) illustrates the E-LMS Motivation consists of the following process worked on this online platform.

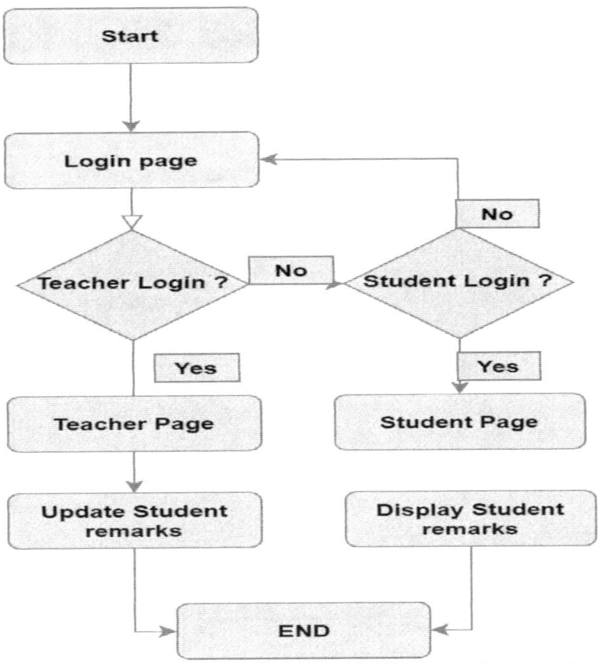

Figure 2. A flowchart of how the proposed E-LMS motivation platform would work.

4. Results and Discussion

For this project, the design of the E-LMS Motivation was defined. The tool aims to collect and track students' performance to provide a visual overview of the relationship between the interaction and the student's performance on the E-LMS, assisting professors or tutors in tracking distance learning students' performance. eight accredited subjects were selected that were taught in the electrical engineering department at Wasit University and 20 students per professor. A special interface has been created for each professor and student, to be accessed via email and a special password for each one. Figure (3) shows the private e-mail addresses and initial passwords for each professor and student.

No.	User	Email	Password
1	Medical	medical@wasit.com	medical
2	Electrical	electrical@wasit.com	electrical
3	Control System Teacher	contorlsystem@wasit.com	controlsystem
4	Advanced power system Teacher	powersystem@wasit.com	powersystem
5	Electrical Machin Teacher	electricalmachines@wasit.com	electricalmachines
6	Electronics Teacher	electronics@wasit.com	electronics
7	Computer Network Teacher	computernetwork@wasit.com	computernetwork
8	laboratories Teacher	laboratories@wasit.com	laboratories
9	Graduation Projects Teacher	graduationproject@wasit.com	graduationproject
10	Communication System teacher	communicationsystem@wasit.com	communicationsystem
11	student1	student1@wasit.com	student1
12	student2	student2@wasit.com	student2
13	student3	student3@wasit.com	student3
14	student4	student4@wasit.com	student4
15	student5	student5@wasit.com	student5
16	student6	student6@wasit.com	student6
17	student7	student7@wasit.com	student7
18	student8	student8@wasit.com	student8
19	student9	student9@wasit.com	student9
20	student10	student10@wasit.com	student10
21	student11	student11@wasit.com	student11
22	student12	student12@wasit.com	student12
23	student13	student13@wasit.com	student13
24	student14	student14@wasit.com	student14
25	student15	student15@wasit.com	student15
26	student16	student16@wasit.com	student16
27	student17	student17@wasit.com	student17
28	student18	student18@wasit.com	student18
29	student19	student19@wasit.com	student19
30	student20	student20@wasit.com	student20

Figure 3. private e-mail addresses.

When opening his destination, the teacher can enter the grades of the subject for which he is responsible. Figure (4) shows the interface for the teacher of the control system, where the name of the subject was written on the upper left side of the interface for each professor.

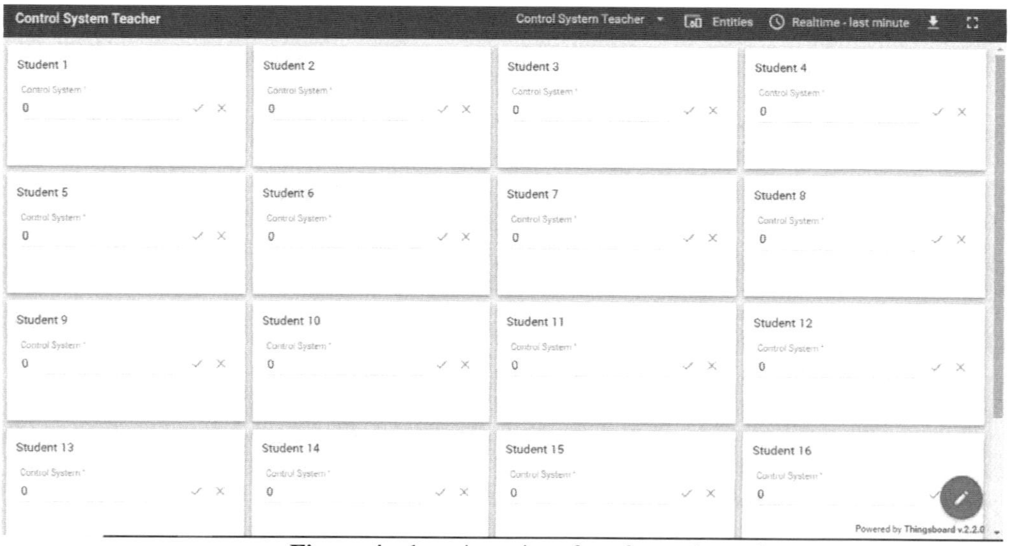

Figure 4. the private interface for each teacher.

As for the student's interface, his academic results that were recorded by the responsible teacher are displayed, as in Figure (5).

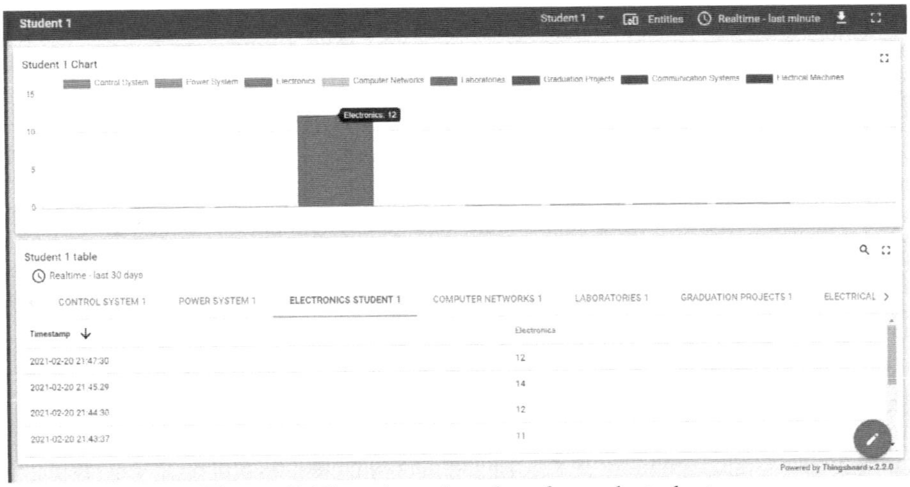

Figure 5. The private interface for each student.

In Figure (5), the interface of each student is divided into two halves. The upper half shows the last grade obtained by the student in each lesson in the form of a graph, while the lower half shows the student's academic results with the date on which the exam was taken. Once the student's guardian obtains the student's e-mail and password, he can see the student's academic level.

5. Conclusion

The Internet of Things (IoT) will grow at a breakneck pace, bringing massive advances in every field with it. When it comes to learning. IoT will carry E-learning to the next level. We can use this to our advantage. IoT's ability to incorporate smart learning, a setting that encourages better learning and increased productivity. Retention rates are important. This innovation in education is significant to develop better people in terms of abilities and knowledge understanding. Distance learning depends on the Internet, so information and communication technologies are used in the e-learning process. E-learning permits students to achieve their classes over a computer network by merging teaching theory and computer network skills. In this study, it was proposed to design an electronic platform in which students' test scores are recorded, as well as their daily attendance by the faculty responsible for them. Where it can clarify the possibility of providing a link between the student and his guardian via the Internet. As a result, parents can play a fundamental role in learning practices by stimulating and improving the student's state of study after monitoring their academic level remotely using a mobile phone, laptop, etc.

References

[1] Faiola A and Matei S A J J o c-m c 2005 Cultural cognitive style and web design: Beyond a behavioral inquiry into computer-mediated communication **11** 375-94

[2] Chen S-Y and Fu Y-C J A 2009 Internet use and academic achievement: Gender differences in early adolescence **44**

[3] Kranz M, Roalter L and Michahelles F 2010 Things that twitter: social networks and the internet of things. In: *What can the Internet of Things do for the Citizen (CIoT) Workshop at The Eighth International Conference on Pervasive Computing (Pervasive 2010),* pp 1-10

[4] Childs S, Blenkinsopp E, Hall A, Walton G J H I and Journal L 2005 Effective e-learning for health professionals and students—barriers and their solutions. A systematic review of the literature—findings from the HeXL project **22** 20-32

[5] Bates A W and Bates T 2005 *Technology, e-learning and distance education*: Psychology Press)

[6] Desai M S, Hart J and Richards T C J E 2008 E-learning: Paradigm shift in education **129**
[7] Mohmmed A O, Khidhir B A, Nazeer A and Vijayan V J J I I S 2020 Emergency remote teaching during Coronavirus pandemic: the current trend and future directive at Middle East College Oman **5** 1-11
[8] Ban Hassan Majeed L F J, Haider Th.Salim Alrikabi 2021 Tactical Thinking and its Relationship with Solving Mathematical Problems Among Mathematics Department Students *International Journal of Emerging Technologies in Learning (iJET)* **16**
[9] Rhema A, Miliszewska I J I i I S and Technology I 2010 Towards e-learning in higher education in Libya **7** 423-37
[10] Duha Khalid Abdul-Rahman Al-Malah H H K J, Haider Th. Salim ALRikabi 2020 Enhancement of educational services by using the internet of things applications for talent and intelligent schools *Periodicals of Engineering and Natural Sciences (PEN)* **8** 2358-66
[11] Chamoso P, González-Briones A, Rodríguez S, Corchado J M J W C and Computing M 2018 Tendencies of technologies and platforms in smart cities: a state-of-the-art review **2018**
[12] Duha Khalid Abdul-Rahman Al-Malah I A A, Haider Th Salim Alrikabi,and Hussain Ali Mutar 2021 Cloud Computing and its Impact on Online Education *IOP Conference Series: Materials Science and Engineering* **1094** 012024
[13] Zubaidi S L, Abdulkareem I H, Hashim K S, Al-Bugharbee H, Ridha H M, Gharghan S K, Al-Qaim F F, Muradov M, Kot P and Al-Khaddar R J W 2020 Hybridised Artificial Neural Network Model with Slime Mould Algorithm: A Novel Methodology for Prediction of Urban Stochastic Water Demand **12** 2692
[14] Zubaidi S L, Al-Bugharbee H, Muhsin Y R, Hashim K and Alkhaddar R 2020 Forecasting of monthly stochastic signal of urban water demand: Baghdad as a case study. In: *IOP Conference Series: Materials Science and Engineering*: IOP Publishing) p 012018
[15] Alaidi A, Yahya O and Alrikabi H 2020 Using Modern Education Technique in Wasit University *International Journal of Interactive Mobile Technologies* **14** 82-94
[16] Alaidi A H M, Yahya O H and AlRikabi H T S 2020 Using modern education technique in Wasit university *International Journal of Interactive Mobile Technologies* **14** 82-94
[17] Shaame A A 2020 Effectiveness of Moodle learning management system for fostering teaching and learning mathematics in secondary schools in Zanzibar. The University of Dodoma)
[18] Madakam S, Lake V, Lake V, Lake V J J o C and Communications 2015 Internet of Things (IoT): A literature review **3** 164
[19] Alseelawi N S, Adnan E K, Hazim H T, Alrikabi H T S and Nasser K W 2020 Design and implementation of an e-learning platform using N-tier architecture *International Journal of Interactive Mobile Technologies* **14** 171-84
[20] Kahrilas G A, Wally L M, Fredrick S J, Hiskey M, Prieto A L, Owens J E J A S C and Engineering 2014 Microwave-assisted green synthesis of silver nanoparticles using orange peel extract **2** 367-76
[21] Al-dabag M, ALRikabi H S and Al-Nima R 2021 Anticipating Atrial Fibrillation Signal Using Efficient Algorithm *International Journal of Online and Biomedical Engineering (iJOE)* **17** 106-20
[22] Mohammed B K, Mortatha M B, Abdalrada A S, ALRikabi H T S J P o E and Sciences N 2021 A comprehensive system for detection of flammable and toxic gases using IoT **9** 702-11
[23] Naseer Ali Hussien A A D A-m, Haider Th.Salim Alrikabi, Faisal Theyab Abed 2021 Monitoring the Consumption of Electrical Energy Based on the Internet of Things Applications *International Journal of Interactive Mobile Technologies (iJIM)* **15**
[24] AjazMoharkan Z, Choudhury T, Gupta S C and Raj G 2017 Internet of Things and its applications in E-learning. In: *2017 3rd International Conference on Computational Intelligence & Communication Technology (CICT)*: IEEE) pp 1-5
[25] Duha Khalid Abdul-Rahman Al-Malah S I H, Haider TH. Salim ALRikabi 2020 The Interactive Role Using the Mozabook Digital Education Application and its Effect on

Enhancing the Performance of eLearning *International Journal of Emerging Technologies in Learning (iJET)* **15** 21-41
[26] Rihab Salah Khairy A S H, Haider TH. Salim ALRikabi2 2021 The Detection of Counterfeit Banknotes Using Ensemble Learning Techniques of
AdaBoost and Voting *International Journal of Intelligent Engineering and Systems* **14** 326-39
[27] Motiwalla L, Tello S J T i and education h 2000 Distance learning on the Internet: An exploratory study **2** 253-64
[28] Pituch K A, Lee Y-k J C and Education 2006 The influence of system characteristics on e-learning use **47** 222-44
[29] Yamada M, Oda T, Matsuo K and Barolli L 2016 Design of an IoT-based e-learning testbed. In: *2016 30th International Conference on Advanced Information Networking and Applications Workshops (WAINA)*: IEEE) pp 720-4
[30] Woolf B P 2010 *Building intelligent interactive tutors: Student-centered strategies for revolutionizing e-learning*: Morgan Kaufmann)
[31] Schumacher C, Ifenthaler D J T I and Education H 2021 Investigating prompts for supporting students' self-regulation–A remaining challenge for learning analytics approaches? **49** 100791

Using IoT Applications for Detection and Monitoring of Underground Cable Fault

Faisal T. Abed*[1], and Haider T. ALRikabi[1]
[1] Department of Electrical Engineering, College of Engineering, Wasit University,Wasit, Iraq

* Corresponding author: ftheyab@uowasit.edu.iq

Abstract. Due to underground environments, wear and tear, rats, etc., underground cables are vulnerable to a large range of faults. It's difficult to diagnose the cause of the fault and the whole cable must be separated from the ground to inspect and repair faults. The project work is intended to use the ESP32 controller for detecting the fault location in underground cable lines from the base station in kilometres. The cable needs to be checked for faults to find a fault in the cable. The basic principle of Ohms law is used in this prototype. The current will vary based on the cable 's fault length. In big cities, instead of overhead wires, the electrical cables run underground. It is hard for detecting the precise location of the fault for the repairing phase of that specific cable if the fault occurs in it underground cable. The suggested device finds the precise the fault position. The system is designed with a group of resistors representing cable length in km and a set of switches are generated to cross check the accurateness of the same at any known distance. The voltage across series resistors changes accordingly in the case of a problem, which can then be transferred to an ADC to produce accurate digital data by using programmed C++ that further displays fault position in the distance. The distance, phase, & time of the fault is shown on a computer or mobile device using the internet. The IoT is used to show information on the internet by using ESP32 WiFi module. The website is generated using HTML coding and the fault information is displayed on the website.

Keywords: Arduino, fault of underground cable, Internet of Things.

1. Introduction

Power distribution networks have been increasingly increasing and their efficiency has been increasing. The whole network is built with a large number of modules that can disturb and interrupt the power supply of the end user[1, 2]. For several globally controlled low voltage- and medium voltage distribution lines, underground cables have been employed for several decades. High-voltage cables in the underground have become ever more used because of the weather, storms, wind, rain and pollution. While the cable production technology is continuously improving, during testing and service, factors still can cause the cables to fail. The Internet of Things (IoT) is the extension of Internet access into daily items and physical devices. A cable that is in good shape will last around 30 years for a lifetime[3-5]. These devices can communicate and interact through the internet with others and are monitor able and remotely controlled, embedded in electronics, Internet networking and other hardware(such as sensors)[6-11]. IOT businesses in a wide range of sectors are using more and more to improve customer experience, improve decision-making and increase the profitability of the organization. IOT companies may also have deeper understanding.

2. Literature Survey

Some experiments and articles were presented and the WSN was discussed. Kevin Ashton, co-originator of MIT's AutoID Center, initially talked about the Internet of Things (IoTs) in a presentation he gave in 1999 to Procter & Gamble (P&G). In 1999, when the stuff beginning with feeling, also emerged in 1999, the new movement of MIT Professor Neil Gershenfeld did not use the precise term but offered simple insights into where IoT was going. Wireless technology, micro-electromechanical systems (MEMS), multiple networks and the Internet has evolved as a result of IoT. This integration helps to break down the silos from OT and IT, enabling analysis of unstructured device-generated data in order to drive information and understanding. While Ashton's had been the primary mention of IoT, since the 1970s the concept of connected devices was built on the Internet and on computer. For example, in the early 1980s, the first Internet appliance had been a coke machine at Carnegie Mellon University. Using the web, programmers can check the status of their machines and determine if a cold drink awaits them, if they wish to make their way to the machine. IoT has been evolved from communication between machines (M2M), specifically from machines that link to one another through a web without human intervention[12-16]. M2M means the connection, handling and collection of data of a computer to the cloud. IoT is a network of trillions of intelligent devices linking individuals, systems and other applications for collecting and exchange data from M2M to the following level. M2M provides a connectivity for IoT as its base. The Internet of things is also a natural extension of SCADA, an application software for process control, collecting data from remote locations in real-time to control equipment. SCADA systems contain elements of the hardware and software. The hardware collects and transmits the data to a device that installs SCADA software, where it is handled and displayed promptly. The SCADA growth is such that SCADA schemes of lately generation have developed into IoT systems of first generation[17-20]. Nevertheless, it was not until mid-2010 that the idea of the IoT ecosystem came into its own. While in part, the Chinese Government announced that it would create IoT an intentional priority in its 5-year plan.

3. Components of the circuit operation Relay

Relays are electric or electronically opening and closing switches. Closing and opening contacts in another circuit monitor the relay on an electric circuit. When a contact with a relay is usually open (NO) as seen in the Relay diagrams, an open contact occurs when the reaction is not triggered. If a relay contact is usually closed (NC), the relay cannot have functioned by a closed contact. If electric current is smeared to the contacts, their condition may change. Relaxes are typically used to turn smaller current in a control circuit and normally do not regulate power consuming units with the exception of small motors and Solenoids with low amplifications. However, relays can "control" larger voltages and amperes with an increasing influence, as a low applied voltage to the relay spiral can result in the contacts switching on a large voltage. Relays can avoid damage to the machinery, including over current, undercurrents, spikes and reverse currents, by detecting electrical anomalies. Relays also are commonly used for heating elements, pilot lights and audible warning systems.

Figure 1. The relay which used in this paper.

3.1. Espressif Systems (ESP32)

It's a 2.4 GHz combo chip for WiFi and Bluetooth with a very powerful 40-nm TSMC technology. It was considered for achieving optimum efficiency and RF, with stable, scalable and reliable applications and power scenarios for many different applications. The ESP32 chip series has been designed for smartphone, wearable, and IoT applications, including ESP32-D0WDQ6, ESP 32-D0WD, ESP32-D1WD and ESP32-S0WD. All advanced features of low-powered chips include fine-grained clock gating, multiple power modes and dynamic power scale. ESP32 is stirred up regularly and just if an identified circumstance has identified in a low-compatibility IoT sensor hub scenario. The low-cost cycle can be employed for minimizing the energy consumed by the chip. A performance of a power amplifier can correspondingly be modified such that the communication field, data rate and power consumption are optimally reconciled. With about 20 external modules, ESP32 can be a fully incorporated solution for Bluetooth, Wi-Fi as well as IoT applications. ESP32 contains an antenna switch, RF-balun, filters, low-noise amplifier, power amplifier and modules for power control. This means that the whole solution covers a small region of the PCB. In addition to the integration of advanced tuning circuits, ESP32 uses CMOS to eliminate external cycle imperfections or adapt to changes in external conditions for a single-chip fully integration of radio and baseline bands. As a result, mass manufacturing of ESP32 solutions requires no costly specialist Wi-Fi test equipment[21-23].

Figure 2. Espressif Systems (ESP32).

3.2. Voltage level converter

You know it can be the challenge, if you try to connect a 3.3V unit to a 5V machine. A tiny system that safely moves down the 5V signal to 3.3V and raises from 3. 3V to 5V by bidirectional CJMCU logic level shifter. This module is available in various 3.3V and 5.0V devices and serves as a bridge to detect the physical conditions that are interested [24].

Figure 3. Voltage level converter.

3.3. Resistance
Resistors with different values were used in the design of this project in accordance with the requirements of the electronic circuit for detecting faults. The values of the resistors used 1Ωand 2 Ω

3.4. Strip Board
The Strip Board is a sheet of electrical current insulation material, which is fitted to electronic elements by welding on one hand with the copper bands, with hole in the strips and the insulation material. The diaphragm plates can be used in various ways, including integrated circuits. The gap between strips and the lumps is 0.1 cm and the hole diameter is 0.04 cm which matches discreet elements. The width is

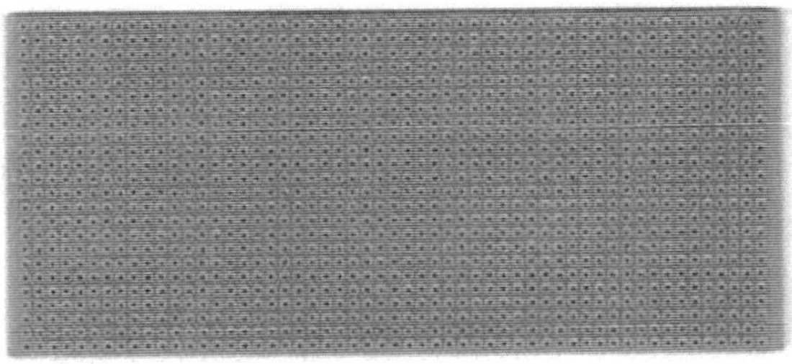

0.15 inches between the strips and the holes, with a diameter of 0.052 inches. Both types in different sizes for small or large circuits can be found on the market.

Figure 4. The Strip Board

4. Types of Faults in Cables
• Open Circuit Fault The cable's open circuit failure is related to when there is a break in the cable's conductor. Megger can control the open circuit fault. To do this, the 3-core cable conductors at the far end are shortened and tiled.

• Short circuit failure: The short circuit defect is called if two conductors of a multi-core cable come into electrical contact because of insulating failure. The two megger terminals are attached to two drivers. If the megger gives zero read, it means that these two conductors have a short circuit fault.
Earth failure: When the cable conductor comes into contact with the surface, the Earth defect is called an earth error or ground fault. The leader is connected to one megger terminal and to Earth the other terminal is connected to detect this error. If Megger demonstrates null reading, the conductors is grounded.

4.1. Actual application of the work

The circuit shown in Figure (5) represents the practical use of the Esp32 project, which represents the control component. The keys indicate the wire extension representing the turning the wire connects to the other wire and the relay is a key component of the circuit.

IOP Conf. Series: Materials Science and Engineering 1184 (2021) 012003 doi:10.1088/1757-899X/1184/1/012003

Figure 5. Software detection of under Ground Cable

4.2. Software detection of under Ground Cable

This model is shown in the figure. 6, the cable control section, in which the form is composed of three cables. There is a certain color in of cable. You should verify this. The four points representing the touch points are included in each cable. As for CA1, CA2, CA3 and CA4 for the first cable seen. In a format which indicates the exact cable interruption date such as the interruption date. The main thing is that the control screen will decide when the cable can be examined. There has been a cutoff of 15 minutes. And there are four points of contact for the possibility of adding any number of cables since this project is not limited to three lines.

4.3. Fault of a Cable

Figure 8 displays the first, second and third wires, each consisting of four conductors. Three wires are shown. In the first cable, a failure takes place at points CA2, CA4, and in the second cable6, CB3 and CB4, as well as in the third cable, CC2 and CC4.Figure 9 indicates the cable break time 21: 54: 05 and the cut date 2019-04-02. It is also integrated into the form where the fault can be handled after it has been identified. The accuracy of the work and so on is what characterizes this project. The time and date of the failures are outlined by its readers as well as the repair time and date.

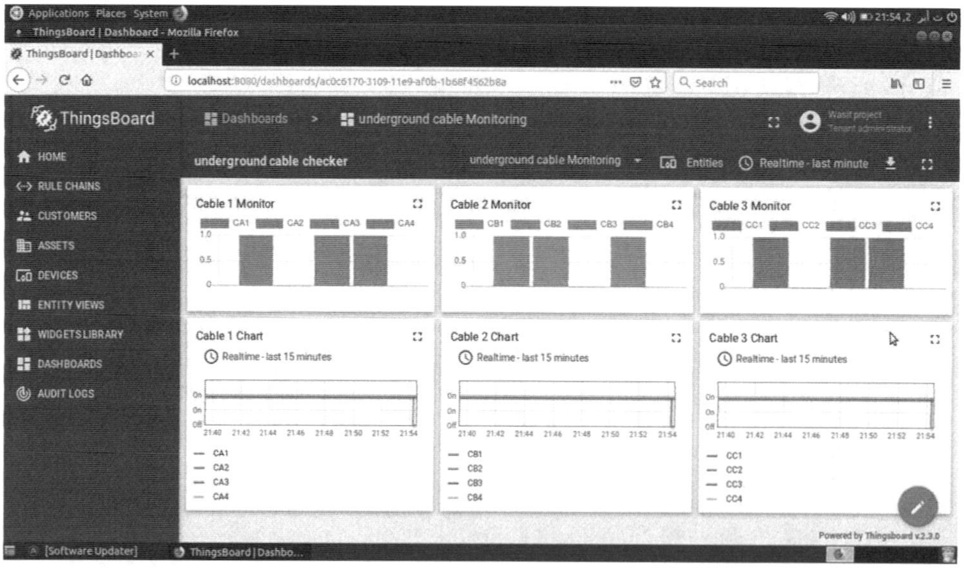

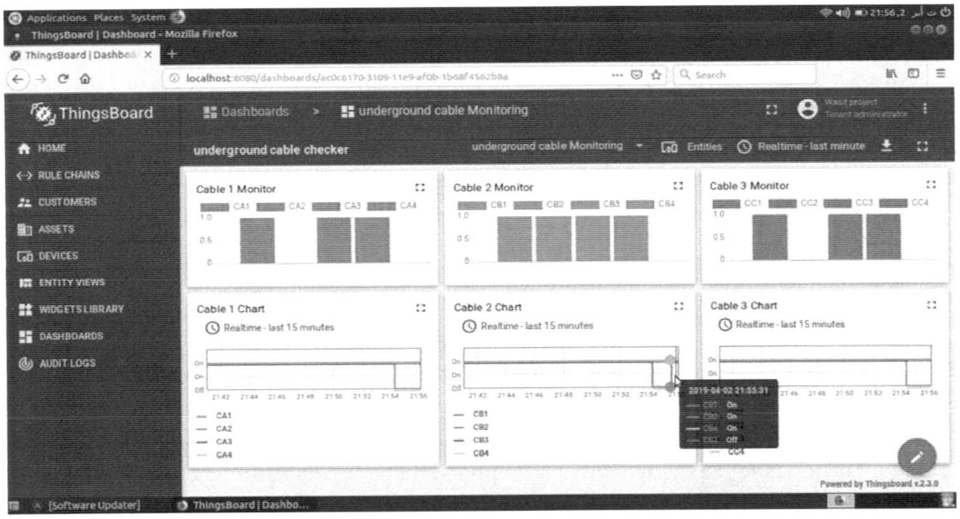

Figure 6. Cable fault detection

5. Conclusions

The works on detecting faults in the cables using the IOT. The electronic circuit containing the Esp32, which acts as a control unit, which contains the Wi-Fi, sends and receives data via the Wi-Fi through the internet, where the circuit is fed by 3V volts and the control voltage On the conversion of the bean the from 3-5V to operate the relay, which acts as an electrical switch to the circuit where in case of a cut in the cable will send a signal to the micro controller in turn sent by the wife to the recipient, either to the computer or mobile and the characteristics of the project was set date and time Unplug the cable and time and date of repair.

References

[1] Das C K, Bass O, Kothapalli G, Mahmoud T S, Habibi D J R and Reviews S E 2018 Overview of energy storage systems in distribution networks: Placement, sizing, operation, and power quality **91** 1205-30

[2] Moreno-Muñoz A 2007 *Power quality: mitigation technologies in a distributed environment*: Springer Science & Business Media)

[3] Borgia E J C C 2014 The Internet of Things vision: Key features, applications and open issues **54** 1-31

[4] Greengard S 2015 *The internet of things*: MIT press)

[5] Zubaidi S L, Al-Bugharbee H, Muhsin Y R, Hashim K and Alkhaddar R 2020 Forecasting of monthly stochastic signal of urban water demand: Baghdad as a case study. In: *IOP Conference Series: Materials Science and Engineering*: IOP Publishing) p 012018

[6] Yahya O H, ALRikabi H T, Al_airaji R a M and Faezipour M 2020 Using Internet of Things Application for Disposing of Solid Waste *International Journal of Interactive Mobile Technologies* **14** 4-18

[7] Patel K K, Patel S M J I j o e s and computing 2016 Internet of things-IOT: definition, characteristics, architecture, enabling technologies, application & future challenges **6**

[8] Mainetti L, Patrono L and Vilei A 2011 Evolution of wireless sensor networks towards the internet of things: A survey. In: *SoftCOM 2011, 19th international conference on software, telecommunications and computer networks*: IEEE) pp 1-6

[9] Hussien N A, Daleh Al-Magsoosi A A, AlRikabi H T and Abed F T J I J o I M T 2021 Monitoring the Consumption of Electrical Energy Based on the Internet of Things Applications **15**

[10] Al-dabag M, ALRikabi H S and Al-Nima R 2021 Anticipating Atrial Fibrillation Signal Using Efficient Algorithm *International Journal of Online and Biomedical Engineering (iJOE)* **17** 106-20
[11] Roa'a M A a, Aljazaery I A, Al_Dulaimi S K, Alrikabi H T S and Informatics 2021 Generation of High Dynamic Range for Enhancing the Panorama Environment *Bulletin of Electrical Engineering* **10**
[12] Montori F, Bedogni L, Di Felice M, Bononi L J P and Computing M 2018 Machine-to-machine wireless communication technologies for the Internet of Things: Taxonomy, comparison and open issues **50** 56-81
[13] Zhou Z, Gong J, He Y and Zhang Y J I C M 2017 Software defined machine-to-machine communication for smart energy management **55** 52-60
[14] Abed F T, ALRikabi H T S and Ibrahim I A 2020 Efficient Energy of Smart Grid Education Models for Modern Electric Power System Engineering in Iraq. In: *IOP Conference Series: Materials Science and Engineering*: IOP Publishing) p 012049
[15] Tuama H, Abbas H, Alseelawi N S and ALRikabi H T S 2020 Bordering a set of energy criteria for the contributing in the transition level to sustainable energy in electrical Iraqi Projects *Periodicals of Engineering and Natural Sciences* **8** 516-25
[16] Zubaidi S L, Abdulkareem I H, Hashim K S, Al-Bugharbee H, Ridha H M, Gharghan S K, Al-Qaim F F, Muradov M, Kot P and Al-Khaddar R J W 2020 Hybridised Artificial Neural Network Model with Slime Mould Algorithm: A Novel Methodology for Prediction of Urban Stochastic Water Demand **12** 2692
[17] Maglaras L A and Jiang J 2014 Intrusion detection in SCADA systems using machine learning techniques. In: *2014 Science and Information Conference*: IEEE) pp 626-31
[18] HAKIM D L, ABDULLAH A G, MULYADI Y J J o E S and Technology 2020 SCADA APPLICATION FOR GEOTHERMAL POWER PLANT **15** 1018-31
[19] AlRikabi H T S, Nasser K W and Alaidi A H M 2020 The application of wireless communication in IOT for saving electrical energy *International Journal of Interactive Mobile Technologies* **14**
[20] Mohammed B K, Mortatha M B, Abdalrada A S, ALRikabi H T S J P o E and Sciences N 2021 A comprehensive system for detection of flammable and toxic gases using IoT **9** 702-11
[21] Hala A. Naman N A H, Mohand Lokman Al-dabag, Haider Th.Salim Alrikabi 2021 Encryption System for Hiding Information Based on Internet of Things *International Journal of Interactive Mobile Technologies (iJIM)* **15**
[22] Alaidi A H M, Aljazaery I A, AlRikabi H T S, Mahmood I N and Abed F T 2020 Design and implementation of a smart traffic light management system controlled wirelessly by arduino *International Journal of Interactive Mobile Technologies* **14** 32-40
[23] Hashim K S, Ewadh H M, Muhsin A A, Zubaidi S L, Kot P, Muradov M, Aljefery M, Al-Khaddar R J W S and Technology 2021 Phosphate removal from water using bottom ash: Adsorption performance, coexisting anions and modelling studies **83** 77-89
[24] Diao F, Li Y, Wang Z, Wu Y and Zhao Y 2020 A Computational Efficient Space-Vector Modulation Scheme for A Hybrid Seven-Level Converter for Medium Voltage Grid-Tied Applications. In: *2020 IEEE Applied Power Electronics Conference and Exposition (APEC)*: IEEE) pp 1786-90

Determination of Point of Zero Charge (PZC) of Concrete Particles Adsorbents

Ebtehal A. Al-Maliky[1], Hatem A. Gzar[1], Mohammed G. Al-Azawy[2]
[1] Civil Engineering Department, Engineering College, Wasit University, Iraq.
[2] Mechanical Engineering Department, Engineering College, Wasit University, Iraq.

Email: ebtehala@uowasit.edu.iq

Abstract. In the present study, the pH point of zero charge (pHpzc) was determined for concrete particles adsorbent. The Mass Titration (MT) Method was adopted, and the adsorbent was used to remove the fluoride from water. The pH of point zero charge of concrete adsorbent was found to be 12. The data of pHpzc of the adsorbents give information about the attraction and repulsion between adsorbents and adsorbates. The mass titration technique was found to be an easy way to determine the pHpzc of the studied material.

Keywords: Point of Zero Charge; Adsorbents; Ionic Strengths; The Mass Titration Method.

1. Introduction

In recent time, waste materials were recognized as cheap adsorbents for the elimination of toxic contaminations. Various types of low-cost adsorbents were tested to examine their ability to remove many kinds of impurities from water and wastewater [1, 2].

It is known that; the sorption mechanisms are challenging to recognize because the surface functional groups and chemical structures are not well identified [3, 4]. Besides, pH has a significant impact on the sorption mechanism since its effects on the contaminant adsorption type in solution. Also, the pH influences the ionization of chemically active sites on the sorbent [5-7].

The pH of point zero charge (PZC) is the value of pH at which the components of surface charge are equal to zero for specified conditions of temperature, pressure, and aqueous solution components[8, 9]. It does not mean that there is no charge at the surface for pH of PZC, but there are equal amounts of both negative and positive charges [10, 11]. Several methods have been developed to determine the PZC in soils and other materials [12, 13]. Three different types of points of zero charge have been detected. The first one is the point zero charge, the second is the point of zero net proton charge, and the third is the point of zero net charge [15 ,14]. The point zero charge is called the isoelectric point because the particles do not move when exposed to an electric field. The method used to measure the PZC is called the salt addition or salt titration method.

The pH of point zero charge is one of the essential features of the adsorbents that greatly influence the percentage of adsorption and is different from one material to another [16-18]. For example, untreated algal biomass with a pHpzc equals to 5 can successfully absorb the highest amount of fluoride at pH = 5 [19-21]. For a different study, the pHpzc of raw and activated biochar of Colocasia esculenta stem has been estimated to be 6.6 and 8.3, respectively. They both attained the maximum removal of fluoride at pH 4 [22, 23].

For the current study, the mass titration method was used to estimate the pHpzc of concrete particles adsorbent collected from demolished buildings. This method was found to be an easy and straightforward method that gives appropriate results.

2. Materials and methods

2.1 Materials

The concrete adsorbent in this study was made from concrete blocks collected from demolished buildings, and it made within specific determinants [24, 25]. Twenty-one clean plastic bottles and orbital shaker (type TS-2, Turki) have been used in the test. The pH of the mixture was adjusted by using 0.1N HCL and/or 0.1N NaOH solution. The pH was measured with a pH meter (Type EZDO model 6011, China). Potassium Chlorate (KCl) with different concentrations was made to be added to the mixture of adsorbent and water to get three ionic strengths.

2.2. Methods

The pH of point zero charge was determined as follows [12, 26]:

- One gram of concrete particles was placed in twenty-one of plastic bottles. The bottles were arranged in three groups, each group include seven bottles.
- The solutions mixture of base, acid and water in the seven bottles of each group is indicated in Table 1.
- the suspensions have been automatically shaken for one hour, and the pH was determined as $pH_i (pH_{water})$.
- Then, one millilitre of (0.1, 1, 2) mol/L of Potassium Chlorate (KCl) was added to each one of the three groups respectively, to get three ionic strengths.
- The bottles were shaken again for 30 minutes.
- The pH values have been measured again as pH_f (pH_{KCl}) and ΔpH was computed as follow $\Delta pH = pH_i \ or \ (pH_{water})$ - $pH_f \ or \ (pH_{KCl})$
- pH_i Versus ΔpH has been plotted.
- The point zero charge (PZC) is the vertical projection of the curve.

TABLE 1. Solutions for PZC determination.

Bottle	Volume of 0.1M of HCl (mL)	Volume of 0.1 M of NaOH (mL)	Volume of Water (mL)
1	5	0	15
2	4	0	16
3	3	0	17
4	2	0	18
5	0	0	20
6	0	3	17
7	0	5	15

3. Results and discussion

The experimental curves (plots of ΔpH vs pH_i) from the salt addition method are shown in Figures 1, 2, and 3. The Results presented the three ionic strengths (2, 1 and 0.1 M) of KCL and pH_{PZC} has been recognized. Figure 1 illustrates the plot of ΔpH vs pH_i for 2M of KCL, and the intersect point is 12, which rpresentes the point of zero charge (see Table 2). For the plots of ΔpH vs pH_i for 1M of KCL, the point of zero charge is 12.1 as shown in Figure 2 and Table3. Moreover, the concentration of 0.1 M of KCL has a 12.04 PZC shown in Figure 3 and Table 4.

TABLE2. Data of determine of PZC of 2M KCI.

Bottles	PH_i (pH water)	pH_f (PH KCl)	Δ pH
1	5.7	8.1	2.4
2	6.3	9.36	3.06
3	7.7	9.55	2.45
4	8	10.03	2.03
5	11.75	12	0.25
6	12.0	12.0	0
7	12.93	12.83	- 0.13

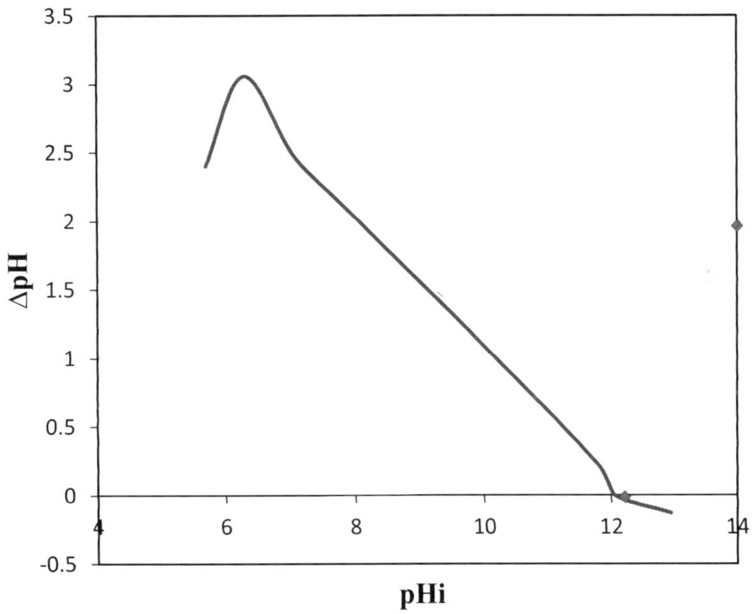

FIGURE 1. PZC Determination Diagram for 2M KCI.

Table 3. Data of determine of PZC of 1M KCI.

Bottles	PH_i (pH $_{water}$)	pH_f (PH $_{KCl}$)	Δ pH
1	6.25	8.5	2.25
2	6.71	9.75	3.04
3	8.3	10.3	2
4	10.7	11.1	0.4
5	11.9	11.94	0.04
6	12.05	12.05	0
7	12.81	12.7	-0.11

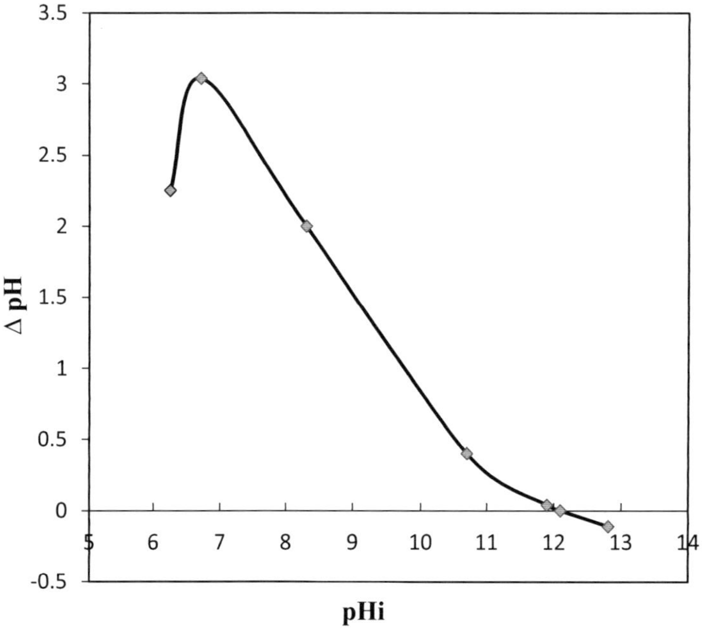

Figure 2. PZC Determination Diagram for 0.1M KCI.

TABLE 4. Data of determine of PZC of 0.1M KCI.

Bottles	PH_i (pH $_{water}$)	pH_f (PH $_{KCl}$)	Δ pH
1	6.25	8.5	2.25
2	6.71	9.75	3.04
3	8.3	10.3	2
4	10.7	11.1	0.4
5	11.9	11.94	0.04
6	12.04	12.04	0
7	12.81	12.7	-0.11

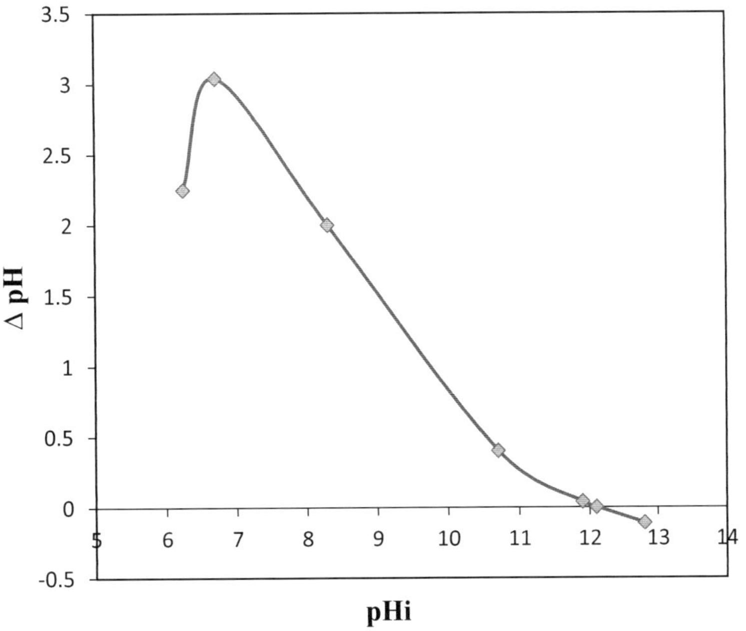

Figure 3. PZC Determination Diagram for 0.1M KCI.

Fluoride adsorption onto concrete particles depended on the pH of the solution as shown in Figure 4, which showed that fluoride uptake increased with the increase of pH of the solution and reached the maximum at pH 6, then decreased when the solution was the same pH value increased. This fact can be explained in terms of the pH point zero charger (pH_{PZC}). The pH_{PZC} of concrete particles is found to be 12, as provided above. The pH_{PZC} means that the net charge on the adsorbent surface is zero. Once the pH of the solution is less than pH_{PZC}, the net surface charge on the solid surface for the adsorbent becomes positive because of the adsorption of excess H^+ which is favoured for fluoride adsorption due to coulombic attraction. On the other hand, the net surface charge is negative because of the desorption of $H^{+,}$ and the fluoride uptake must compete with coulombic repulsion when the pH of the mixture is bigger than pH_{PZC}. For this reason, a proper removal of fluoride below pH_{PZC} can be noticed due to the presence of positively charged sites on the adsorbent surface.

It can be seen that the amount of fluoride uptake is reduced in low pH even so the pH of the solution is below the pHPZC, which can be ascribed to the company of weak hydrofluoric acid, which lowered the free fluoride ions availability to be adsorbed. Moreover, it can be notched a drop in fluoride uptake in the alkaline values even if the surface has a positive charge. This behaviour ascribes the fluoride and hydroxyl ions intense competition to bind onto active sites. It is noteworthy that the final mixture pH (after adsorption) is near to 12 for initial pH of (3 to 11), which indicates a presence of a large number of exchangeable OH^-. These results are consistent with the studies conducted by Raichur, A.M. and M. Jyoti Basu [27] and Tor, A. [28].

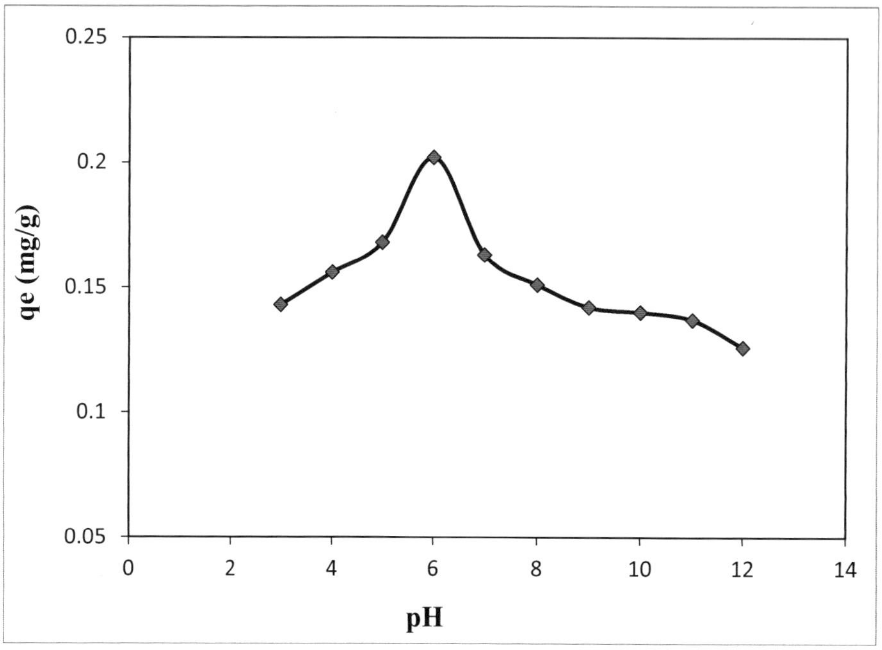

Figure 4. Effect of pH value on fluoride uptake.

4. Conclusion

Knowing the point of zero charge of the adsorbent affords information for the attraction and repulsion between adsorbent and adsorbate. For the current study, the mass titration method was used to estimate the pHpzc of concrete particles adsorbent collected from demolished buildings

The results indicate that the final mixture pH (after adsorption) is near to 12 for initial pH of (3 to 11), which indicates a presence of a large number of exchangeable OH−. It was also found that the mas titration method is suitable and straightforward to calculate the point of zero charge of concrete particles adsorbent. The calculations were achieved in a short time, and they can be evaluated in a typical laboratory session.

References

[1] Sabreen L. kareem, Sohaib K. Al-Mamoori, Laheab A. AL-MALIKI, Mohammed Q. AL-Dulaimi and AL-ANSARI N 2020 Optimum Location For Landfills Landfill Site Selection Using Gis Technique: Al-Naja City As A Case Study *Cogent Engineering*

[2] De Gisi S, Lofrano G, Grassi M and Notarnicola M 2016 Characteristics and adsorption capacities of low-cost sorbents for wastewater treatment: A review *Sustainable Materials and Technologies* **9** 10-40

[3] Lei C, Wang C, Chen W, He M and Huang B 2020 Polyaniline@magnetic chitosan nanomaterials for highly efficient simultaneous adsorption and in-situ chemical reduction of hexavalent chromium: Removal efficacy and mechanisms *Science of The Total Environment* **733** 139316

[4] Hamid Rasool M, Al-Maliki L A, Al-Mamoori S K and Al-Ansari N 2021 Estimation of Uplift Pressure Equation at Key Points under Floor of Hydraulic Structures *Cogent Engineering* **8** 1917287

[5] Fiol N and Villaescusa I 2009 Determination of sorbent point zero charge: usefulness in sorption studies *Environmental Chemistry Letters* **7** 79-84

[6] Khoshnam F, Zargar B and Moghadam M R 2019 Adsorption and removal of ametryn using graphene oxide nano-sheets from farm waste water and optimization using response surface methodology *Journal of the Iranian Chemical Society* **16** 1383-90
[7] Al-Maliki L A, Farhan S L, Jasim I A, Al-Mamoori S K and Al-Ansari N 2021 Perceptions about water pollution among university students: A case study from Iraq *Cogent Engineering* **8** 1895473
[8] Kosmulski M 2020 The pH dependent surface charging and points of zero charge. VIII. Update *Advances in Colloid and Interface Science* **275** 102064
[9] Jasim I A, Mahmood T S, Al-Mamoori S K and Al-Maliki L A 2021 The relationship between traffic congestion and land uses: A case study of Al-Kut city, Iraq *Journal of Urban Regeneration & Renewal/ Henry Stewart Publications* **14** 264-71
[10] Bakatula E N, Richard D, Neculita C M and Zagury G J 2018 Determination of point of zero charge of natural organic materials *Environmental Science and Pollution Research* **25** 7823-33
[11] Al-Mamoori S K, Kareem S L, Al-Maliki L A and El-Tawil K 2020 Geotechnical Maps for Angle of Internal Friction of An-Najaf Soil-Iraq Using GIS *Wasit Journal of Engineering Sciences* **8**
[12] Durán A C, Flores I, Perozo C and Pernalete Z 2006 Immobilization of lead by a vermicompost and its effect on white bean (Vigna Sinenis var. Apure) uptake *International Journal of Environmental Science & Technology* **3** 203-12
[13] Al-Mamoori S K and Al-Maliki L A 2016 EVALUATION OF SUITABILITY OF DRAINAGE WATER OF AL-HUSSAINIA SECTOR (KUT IRAQ) TO IRRIGATE COTTON CROP *Kufa Journal of Engineering* **7** 67-78
[14] Hao W, Flynn S L, Alessi D S and Konhauser K O 2018 Change of the point of zero net proton charge (pHPZNPC) of clay minerals with ionic strength *Chemical Geology* **493** 458-67
[15] Mohammad O I and Al-Maliki L A 2014 Evaluation of Suitability of Drainage Water of Al-Hussainia sector (Kut-Iraq) For Irrigation *Wasit Journal of Engineering Sciences* **2** 30-45
[16] Yang T, Xu Y, Huang Q, Sun Y, Liang X, Wang L, Qin X and Zhao L 2021 Adsorption characteristics and the removal mechanism of two novel Fe-Zn composite modified biochar for Cd(II) in water *Bioresource Technology* **333** 125078
[17] Al-Maliki L A J, Al-Mamoori S K, El-Tawel K, Hussain H M, Al-Ansari N and Jawad Al Ali M 2018 Bearing Capacity Map for An-Najaf and Kufa Cities Using GIS *Engineering* **10** 262-9
[18] Al-Mamoori S K, Attiyah A N, Al-Maliki L A, Al-Sulttani A H, El-Tawil K and Hussain H M 2020 *Modern Applications of Geotechnical Engineering and Construction*: Springer) pp 329-48
[19] Mukherjee S, Mondal M, Banerjee S and Halder G 2017 Elucidation of the sorptive uptake of fluoride by Ca2+-treated and untreated algal biomass of Nostoc sp. (BTA394): Isotherm, kinetics, thermodynamics and safe disposal *Process Safety and Environmental Protection* **107** 334-45
[20] Al-Mamoori S K, Al-Maliki L A J, El-Tawel K, Hussain H M, Al-Ansari N and Al Ali M J 2019 Chloride, Calcium Carbonate and Total Soluble Salts Contents Distribution for An-Najaf and Al-Kufa Cities' Soil by Using GIS *Geotechnical and Geological Engineering* **37** 2207-25
[21] Al-Mamoori S K, Al-Maliki L A J, Al-Sulttani A H, El-Tawil K, Hussain H M and Al-Ansari N 2020 Horizontal and Vertical Geotechnical Variations of Soils According to USCS Classification for the City of An-Najaf, Iraq Using GIS *Geotechnical and Geological Engineering* **38** 1919-38
[22] Mukherjee S and Halder G 2016 Assessment of fluoride uptake performance of raw biomass and activated biochar of Colocasia esculenta stem: optimization through response surface methodology *Environmental Progress & Sustainable Energy* **35** 1305-16
[23] Jasim I A, Farhan S L, Al-Maliki L A and Al-Mamoori S K 2021 Climatic Treatments for Housing in the Traditional Holy Cities: A Comparison between Najaf and Yazd Cities *IOP Conference Series: Earth and Environmental Science* **754** 012017
[24] Almaliky E A and Gzar H A 2020 Geomaterials as Cost Effective Sorbent to Remove Fluoride from Water *Key Engineering Materials* **870** 107-21
[25] Al-Shammaa A K M and Al-Mamoori S K 2021 A Design Chart for Optimum Dimensions of Reinforced Concrete Cantilever Beams *IOP Conference Series: Materials Science and Engineering* **1094** 012023

[26] Al-Mamoori S K, Al-Maliki L A, Hussain H M and Al-Ali M J 2018 Distribution of Sulfate Content and Organic Matter in An-Najaf and Al-Kufa Cities'soil Using Gis *KUFA JOURNAL OF ENGINEERING* **9** 92-111

[27] Raichur A M and Jyoti Basu M 2001 Adsorption of fluoride onto mixed rare earth oxides *Separation and Purification Technology* **24** 121-7

[28] Tor A 2006 Removal of fluoride from an aqueous solution by using montmorillonite *Desalination* **201** 267-76

IOP Conf. Series: Materials Science and Engineering 1184 (2021) 012005 doi:10.1088/1757-899X/1184/1/012005

Concrete particles for fluoride removal using continued fixed-bed and fluidized-bed systems

Hatem A. Gzar[1,*], Ebtehal A. Al-Maliky[1] and Mohammed G. Al-Azawy[2]
[1] Civil Engineering Department, Engineering College, Wasit University, Iraq.

Email: hatemasal@uowasit.edu.iq

Abstract. In the present study, fluoride ions removal from aqueous solution was studied by adsorption using a new low cost adsorbent which is concrete particles in continuous systems. The continuous systems included fixed and fluidized bed reactor. several operating conditions were used for both of fixed and fluidized column; flow rate (6,9 and 12) l/h, bed depth (6,12,20) cm inlet fluoride concentration (10,15 and 20) mg/l. Thomas and Yoon–Nelson models were used in this study using the data of breakthrough curves for the optimal conditions (bed height of 20 cm, influent flow rate of 6 l/h and inlet fluoride concentration of 10 mg/l). Adsorption capacity of the concrete particles was found to be 0.818 mg/g and 0.73 mg/g for fluidized and fixed bed respectively for the optimum conditions. The adsorption test using concrete particles adsorbent revealed the good performance of removing fluoride from aqueous solution.

Keywords: Concrete particles, Adsorption, Fluidized and Fixed-bed column, Fluoride removal, water treatment.

1. Introduction

Fluoride contamination is common in many countries around the world. Environmentally, there are two different routes responsible for fluoride contamination; anthropogenic sources and natural sources [1]. The first source can be resulted from several industrial activities, while the natural source of fluoride in water comes from the rocks that containing fluoride through which the water has infiltrated [2]. Fluoride can be useful or harmful and the allowable limit of fluoride is between (1mg/l-1.5mg/l) as establish by the World Health Organization (WHO) [3]. Fluoride can cause acute health problems when the concentration exceeds 1.5 mg/l such as dental and skeletal fluorosis, non-skeletal fluorosis.
Furthermore, excess fluoride consumption for long time causes serious diseases like cancer, alzheimer, kidney damage, etc. For more than 30 nations over the world, the fluorosis was widespread among 200 million groups of people [4]. Several methods have been used to eliminate the fluoride ions in aqueous solution, the main weaknesses of many of these methods are the maintenance and operational high costs, small removal capacity, unwanted effects on water quality, complexity of the treatment procedures and production of big amount of sludge [5].

Among the techniques that used to treat fluoride in water the adsorption has been broadly used and showed good results and it is found to be more attractive technique to remove fluoride in term of low cost and simple in design and operation [6]. A lots of materials from low cost type were used as an adsorbent. Generally, the low-cost substance that can be used as adsorbents are available in local places and need the least processing before using [7]. It has been evaluated such materials to remove fluoride from water, like light weight concrete [8] and cement paste [9] alumina cement granules [10], and they was showed good result to treat excess concentrations of fluoride. On the other side, the big amounts of demolished concrete generated from deteriorated and obsolete structures produces acute ecological and environmental problem. Concrete is a building material that produced from mixing of different materials like cement, sand, aggregate and water. The major advantage of using concrete waste material as an adsorbent for fluoride removal over the other chemical treatment techniques is that it does not create any chemical sludge. In the present study, concrete particles from local demolished buildings has been used as low cost adsorbent to remove excess concentrations of fluoride ions from water.

Continuous fixed bed system was used in the current study. Two important goals could be achieved simultaneously by using concrete particles as an adsorbent, recycle concrete waste and reduce the excess fluoride concentrations in water or wastewater, thus reducing the percentage of contaminants around us.

2. Materials and methods

2.1 Materials

2.1.1Collection and Preparation of the Adsorbent. The adsorbent was made from concrete blocks from demolished buildings. The blocks were collected from two different local places they were cleaned, crashed with hand hammer to get small portions and grinded many times by mill agate mortar then sieved to obtain particles sizes between (0.6-0.075) mm. These particles have been washed with tap water several times then rinsed with distilled water. The product was dried in the oven for two days under 105°C then the adsorbent have been kept in plastic container to use in the experiments.

2.1.2 Fluoride Solution. The used chemicals were of analytical grade, and deionized water was used for solution preparation. Fluoride stock solution was prepared by dissolving 0.22 g of NaF in 1000 ml of deionized water and appropriate dilution from the stock solution was done to make fluoride solutions with a specific concentration for the experiments. The pH of the mixture was adjusted by using 0.1N HCL and/or 0.1N NaOH solution. The pH was measured with pH meter (Type EZDO model 6011,China). Before the experiments, all the polypropylene ware used for experimentation, dilution and storage were cleaned with detergent, thoroughly rinsed with tap water and finally rinsed with distilled water before use.

2.2. Methods

2.2.1. Continuous system experiment. The continuous system consists of two glass columns with 50 mm internal diameter and height of 1000 mm and 500 mm for fluidized and fixed columns respectively. Also, there is plastic distributors at the top and bottom for both of fixed and fluidized bed column to support the concrete particles and to remove the presence of particles in the effluent. A selected bed depth from concrete particles was laid in the two columns. The solution of fluoride ions was made in the feed basin of 50 L capacity using distilled water with a known concentration. The solution was adjusted using 0.1N NaOH or HCl. The solution was pumped to the adsorption columns through the flow meters with desired flow rate. It has been designated two points for samples taken.

Ones from the bottom of fixed bed and the other from the top of the fluidized bed column. Samples were taken from time to time and the concentration was accounted using spectrophotometer method.

The experiments have been done at 30°C. The breakthrough curves were formed from the drawing of discharge concentration (C_e/C_0) versus time.

2.2.2. Breakthrough curves models. The column breakthrough curves were formed from the drawing of discharge concentration (C_e/C_0) versus time (t), where C_0 is the initial concentration (mg/l), C_e is the concentration of fluoride at equilibrium (mg/l). Two analytical equations were depended which are

Thomas and Yoon-Nelson models to explain the S-shaped curve of breakthrough after established the best conditions. The linear equation of Thomas model can be written as [11]:

$$\ln\left(\frac{C_0}{C_e} - 1\right) = \frac{K_{th} q_0 m}{Q} - K_{Th} C_0 t \qquad (1)$$

K_{th} is the constant of Thomas model (L/min.mg), q_0 is the adsorption capacity of the bed (mg/g) and t is the time of total flow (min), m is the amount of adsorbent in the column (g), C_0 is the inlet of fluoride concentration (mg/l), C_t is fluoride discharge concentration (mg/l) and Q is the flow rate (ml/min). The linear drawing of $\ln\left(\frac{C_0}{C_t} - 1\right)$ against t can be provided the values of K_{Th} and q_o [12].

The linear equation of Yoon Nelson model as below [13]:

$$\ln\left(\frac{C_t}{C_0 - C_t}\right) = K_{YN} t - \tau K_{YN} \qquad (2)$$

Where: τ (min) is the time necessary for 50% adsorbate breakthrough which determined from the intercept of linear draw of ln (C_t/C_o-C_t) $\ln\left(\frac{C_t}{C_0 - C_t}\right)$ against t, while the slope represents the rate constant K_{YN} (min^{-1}) [14]. The experimental conditions for the continuous system are illustrated in table 1.

Table 1. Experimental conditions for the continuous system.

Adsorbent	Concrete particles material
Adsorbate	F^-
Bed Depth (cm)	6, 12, 20
Flow rate (l/h)	6, 9, 12
Initial Concentration (mg/l)	10,15,20
Temperature,°C	30
Particle size (mm)	0.075 - 0.6
pH	6.00

3. Results and discussion

3.1. Continuous system

The fixed and fluidized bed column experiment of has been obtained at 30°C and pH6. The effect of the pH value of the solution has been studied previous in batch system [15]. The breakthrough curves for both columns at different flow rates (6, 9, 12 l/h) are shown in figures 1 and 2.

It can be seen that the breakthrough curve slope for the fixed and fluidized beds increased with the increasing of the flow rate of the fluid. This result can explained by decreasing in contact time between the fluoride ions and concrete particles in the bed for the greater velocity which mean there was not enough time to achieve the equilibrium. These results are also consistent with that found by [16].

It has been used three different bed depths of concrete adsorbent (6, 12 and 20) cm. The results for both columns are illustrated in figures 3 and 4. It can notice that when the bed depth is smaller, the concentration of effluent fluoride increased rapidly comparing with bigger depth. As well as the bigger bed needs more time to be saturated since there is more surface area from the adsorbent that offering additional numbers of binding sites for fluoride ions to be adsorbed. These conclusions are same with [17].

The effects of different concentrations of fluoride ions are presented in figures 5 and 6. The results showed that the adsorption rate decreased when the initial concentration decreased and the breakthrough time was longer when fluoride concentration was low. This behaviour can be ascribed to diffusion rate that controlling by the concentration gradient. The increasing in fluoride concentration make the

IOP Conf. Series: Materials Science and Engineering 1184 (2021) 012005 doi:10.1088/1757-899X/1184/1/012005

breakthrough time and exhaustion time smaller as the binding sites will be exhausted fast. These results are similar to that obtained by [18].

It can noticed that the effectiveness of fluidized bed column is slightly better than fixed bed column and the breakthrough time in the fluidized bed is longer than the fixed bed for the same operation conditions. this behaviour can explained by several reasons such as the large surface area of the particle, the high mass transfer rate, the rapid mixing process in fluidized column and the occurrence of dead zone.

3.1. Modelling of continuous system

After set the best removal conditions (L=20 cm, Q=6 l/h, C0=10 mg/l), Thomas and Yoon -Nelson models have been used to predict the breakthrough curves for both column fixed and fluidized bed.

The two models showed a good match with the experimental data as illustrated in figures 7,8,9 and 10. In fixed bed column, Yoon and Nelson model (R^2=0.9812) gives a better-fitting compared to Thomas model (R^2=0.975). In fluidized bed, Thomas model with R^2=0.9724 gives a little better fitting compared to Yoon-Nelson model with R^2=0.9718. This result may imply that the internal and external diffusions were not the only limiting step.

The values of R^2 for both models are revealing the ability of using both models to expect the performance of adsorption for fluoride adsorption onto concrete particles in fixed and fluidized bed column these results are same with [13] [19].

Adsorption capacity of the concrete particles was found to be 0.818 mg/g and 0.73 mg/g for fluidized and fixed bed respectively for the optimum conditions

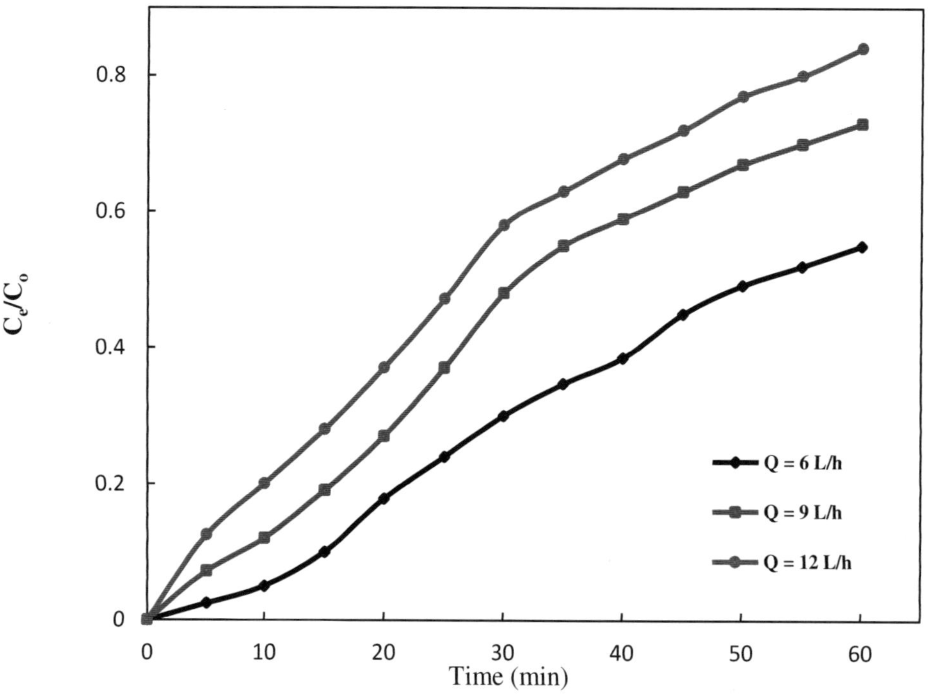

Figure 1. Breakthrough curves of fixed bed column for fluoride ions adsorption onto concrete particles at different flow rates at 30°C (pH 6, C0=15 mg/l, and L=12 cm).

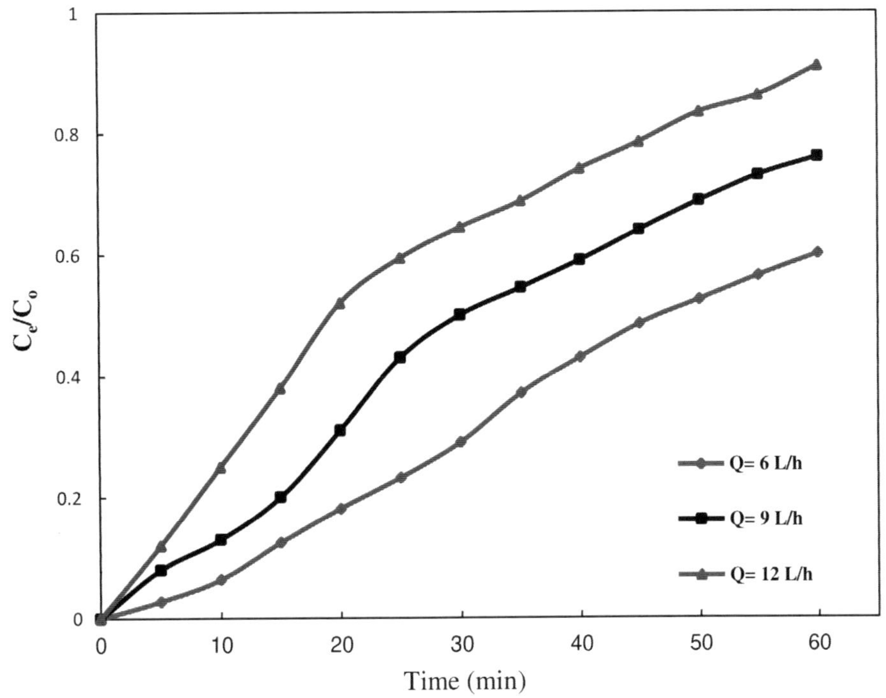

Figure 2. Breakthrough curves of fluidized bed column for fluoride ions adsorption onto concrete particles at different flow rates at 30°C (pH 6, C_0=15 mg/l, and L=12 cm).

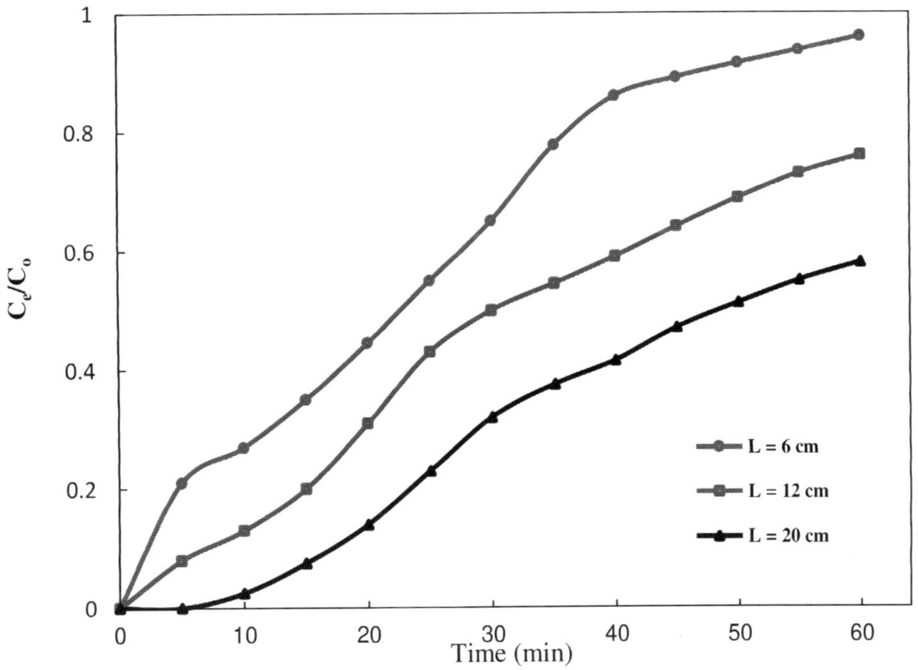

Figure 3. Breakthrough curves of fixed bed column of fluoride ions adsorption onto concrete particles for different bed depths at 30°C (pH 6, C_0=15 mg/L, and Q=9L/h).

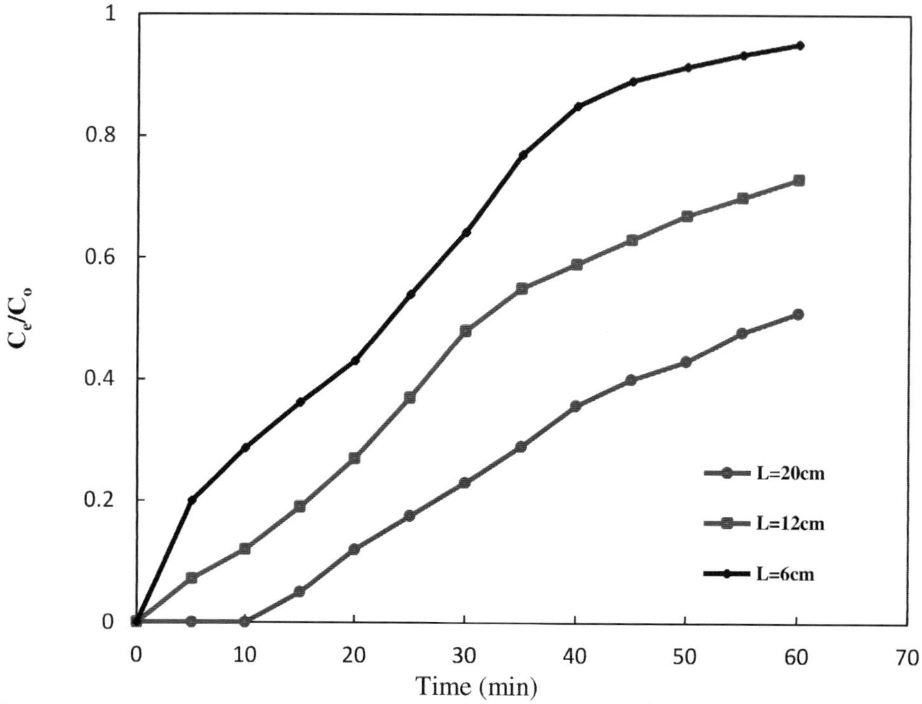

Figure 4. Breakthrough curves of fluidized bed column of fluoride ions adsorption onto concrete particles for different bed depths at 30°C (pH 6, C_0=15 mg/L, and Q=9L/h).

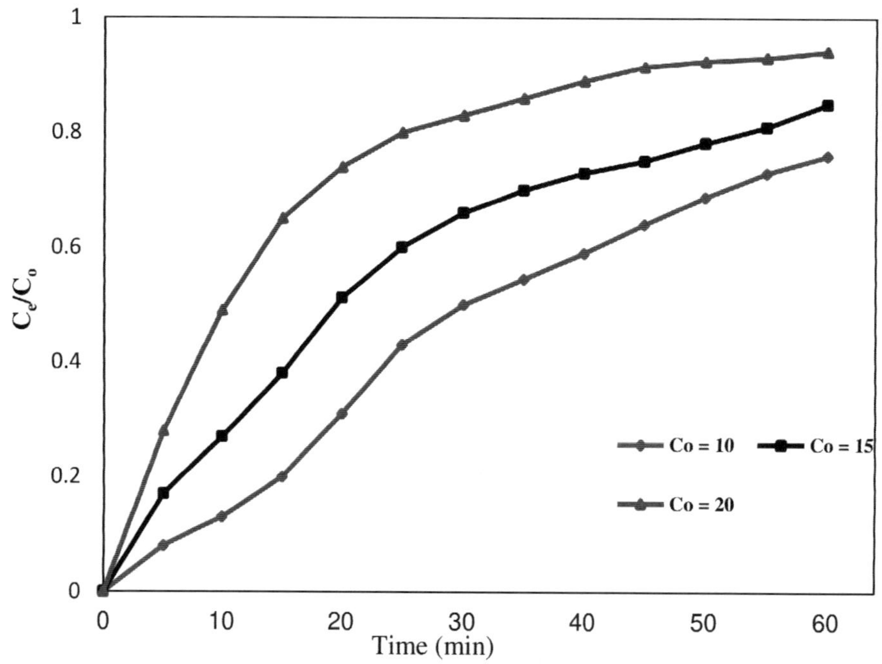

Figure 5. Breakthrough curves of fixed bed column for fluoride ions adsorption onto concrete particles at different initial concentrations at 30°C (pH6,Q= 9 mg/l, L=12cm).

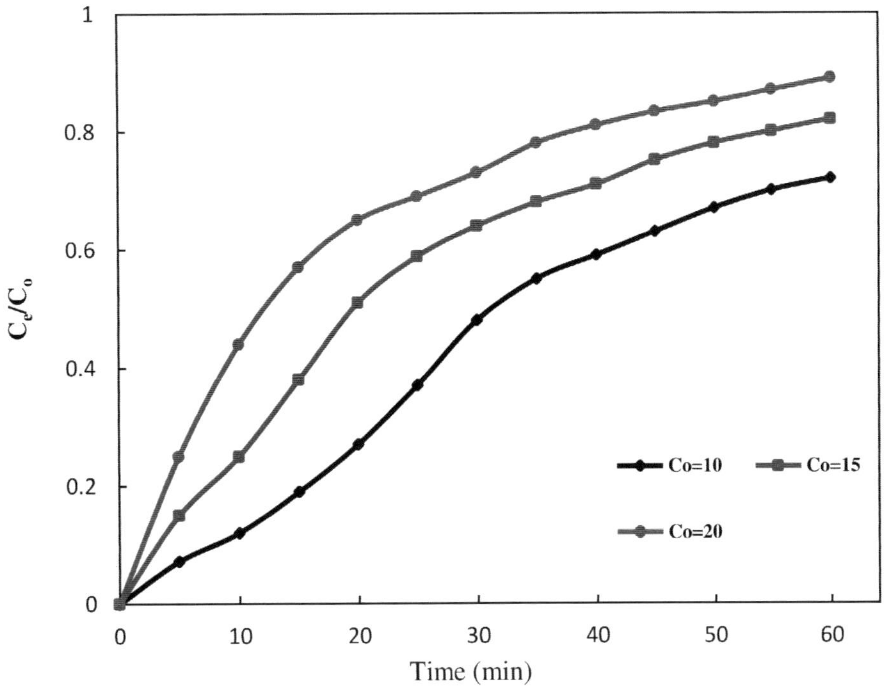

Figure 6. Breakthrough curves of fluidized bed column for fluoride ions adsorption onto concrete particles at different initial concentrations at 30°C (pH6,Q= 9 mg/l, L=12cm).

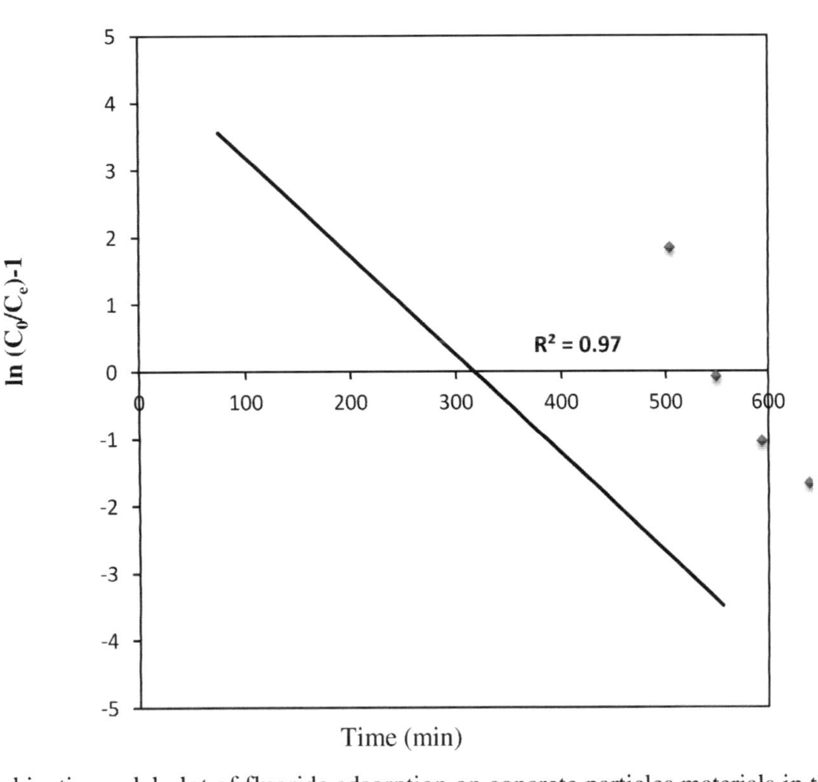

Figure 7. Thomas kinetic model plot of fluoride adsorption on concrete particles materials in the fixed bed.

IOP Conf. Series: Materials Science and Engineering 1184 (2021) 012005 doi:10.1088/1757-899X/1184/1/012005

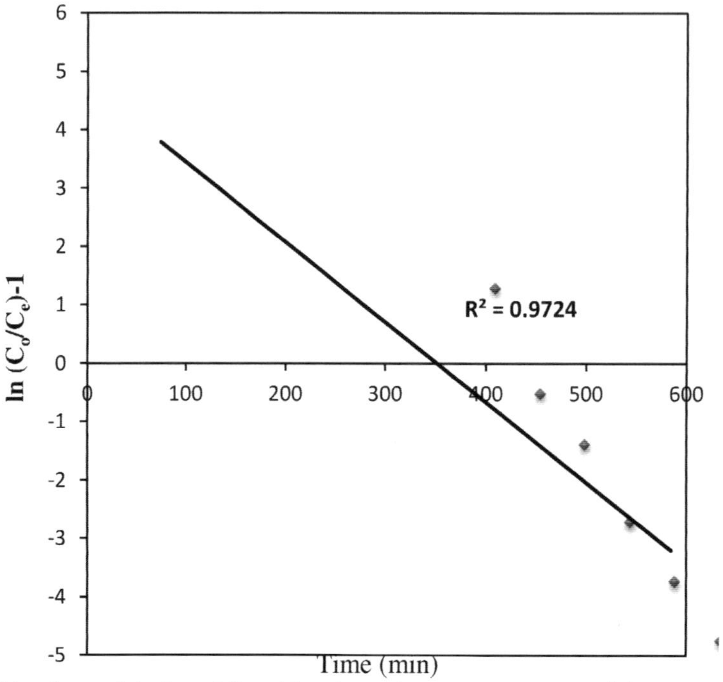

Figure 8. Thomas kinetic model plot of fluoride adsorption on concrete particles materials in the fluidized bed.

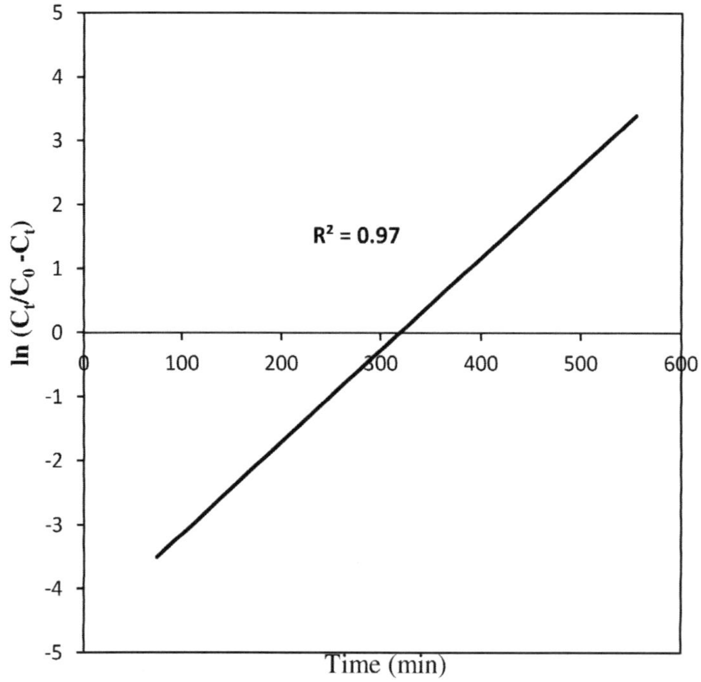

Figure 9. Yoon and Nelson model plot of fluoride adsorption on concrete particles materials in the fluidized bed.

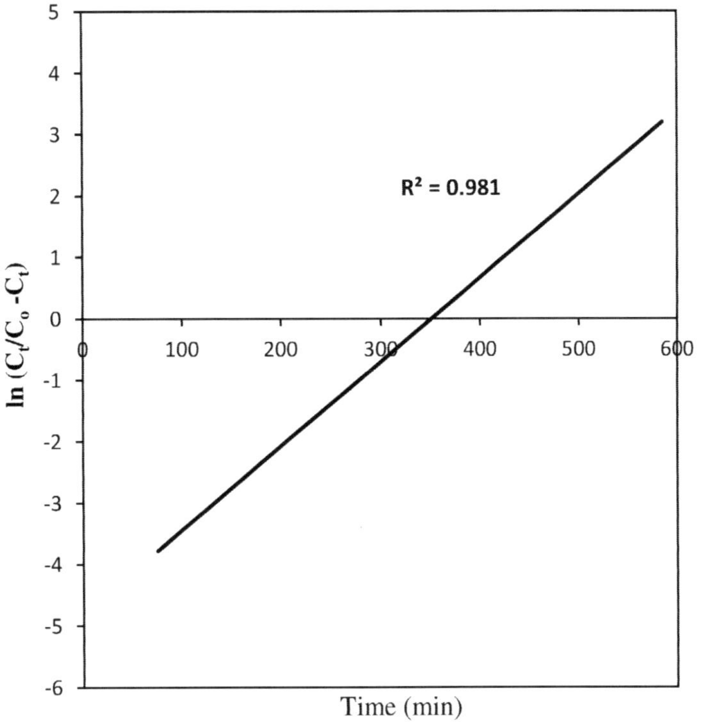

Figure 10. Yoon and Nelson model plot of fluoride adsorption on concrete particles materials in the fixed bed.

4. Conclusion

The current study shows that the ability of concrete adsorbent to remove fluoride ions in water. The performance of concrete particles in continuous mode and the breakthrough time were influenced by variation in bed depth, flow rate along with initial fluoride concentration. The column adsorption was enhanced by the decreased in flow rate and increase in bed depth. The experimental data for the optimum conditions successfully applied to the mathematical models of Thomas and Yoon-Nelson which were fitted well with these models.

5. References

[1] Samatya S, Yuksel U, Yuksel M and Kabay N 2007 Removal of Fluoride from Water by Metal Ions (Al3+, La3+and ZrO) Loaded Natural Zeolite *Sep. Sci. Technol.* **42** 2033–2047

[2] Singh K, Lataye D H and Wasewar K L 2016 Removal of Fluoride from Aqueous Solution by using Bael (Aegle Marmelos) Shell Activated Carbon: Kinetic, Equilibrium and Thermodynamic study *Fluor. Chem*

[3] Das N, Pattanaik P and Das R 2005 Defluoridation of drinking water using activated titanium rich bauxite *J.Colloid and Interface Sci.* **292** 1–10

[4] Mukherjee S and Halder G 2018 A review on the sorptive elimination of fluoride from contaminated wastewater *J. Environ. Chem. Eng.* **6** 1257–1270

[5] Yadav A K, Kaushik C P, Haritash A K, Kansal A and Rani N 2006 Defluoridation of groundwater using brick powder as an adsorbent *J. Hazard. Mater.* **128** 289–293

[6] Bhatnagar A, Kumar E and Sillanpää M 2011 Fluoride removal from water by adsorption-A review *Chem. Eng. J.*, **171** 811–840

[7] De Gisi S, Lofrano G, Grassi M and Notarnicola M 2016 Characteristics and adsorption capacities of low-cost sorbents for wastewater treatment: A review *Sustain. Mater. Technol.* **9** 10–40

[8] Oguz E 2007 Equilibrium isotherms and kinetics studies for the sorption of fluoride on light weight concrete materials *Colloids Surfaces A Physicochem. Eng. Asp.* **295** 258–263
[9] Kang W, Kim E and Park J 2007 Fluoride removal capacity of cement paste *Des.* **202** 38–44
[10] Ayoob S and Gupta A K 2009 Performance evaluation of alumina cement granules in removing fluoride from natural and synthetic waters *Chem. Eng. J.* **150** 485–491
[11] Liua J, Hub C and Huang Q 2019 Adsorption of Cu 2+ , Pb 2+ , and Cd 2+ onto oiltea shell from water *Bioresour. Technol.* **271** 487–491
[12] Han R, Wang Y, Zou W, Wang Y and Shi J 2007 Comparison of linear and nonlinear analysis in estimating the Thomas model parameters for methylene blue adsorption onto natural zeolite in fixed-bed column *J. Hazard. Mater.* **145** 331–335
[13] Talat M, Mohan S, Dixit V, Singh D K, Hasan S H and Srivastava O N 2018 Effective removal of fluoride from water by coconut husk activated carbon in fixed bed column: experimental and breakthrough curves analysis *Groundw. Sustain. Dev.* **7** 48-55
[14] Kundu S and Gupta A K 2007 As(III) removal from aqueous medium in fixed bed using iron oxide-coated cement (IOCC): Experimental and modeling studies," *Chem. Eng. J.* **129** 123–131
[15] Almaliky E A and Gzar H A 2020 Geomaterials as cost effective sorbent to remove fluoride from water *Key Eng. Mater.* **870** 107–121
[16] Ye Y, Yang J, Jiang W, Kang J, Hu Y and Hao H 2018 Fluoride removal from water using a magnesia-pullulan composite in a continuous fi xed-bed column *J. Environ. Manage.* **206** 929–937
[17] Kamble S P, Deshpande G, Barve P P, Rayalu S, Labhsetwar N K, Malyshew A and Kulkarni B D 2010 Adsorption of fluoride from aqueous solution by alumina of alkoxide nature : Batch and continuous operation *DES* **264** 15–23
[18] Rojas-Mayorga C K, Bonilla-Petriciolet A, Sánchez-Ruiz F J, Moreno-Pérez J, Reynel-Ávila H E, Aguayo-Villarreal I A and Mendoza-Castillo D I 2015 Breakthrough curve modeling of liquid-phase adsorption of fluoride ions on aluminum-doped bone char using micro-columns: effectiveness of data fitting approaches *Journal of Molecular Liquids* **208** 114-121
[19] de Franco M A E, de Carvalho C B, Bonetto M M, de Pelegrini Soares R and Féris L A 2018 Diclofenac removal from water by adsorption using activated carbon in batch mode and fixed-bed column: isotherms, thermodynamic study and breakthrough curves modeling *Journal of Cleaner Production* **181** 145-154

Investigating the effects of limestone and rice husk ash on the mechanical and durability properties of concrete

A Abdulridha[1]
[1] Department of Civil Engineering, College of Engineering, University of Warith AL-Anbiya'a, Kerbala, Iraq.
Email: ali.abd@g.uowa.edu.iq

Abstract. Urbanisation and industrial development have led to an increasing need for a new building which in turn increase the demand for raw construction materials such as coarse aggregate. Therefore, researchers focus shifted towards implementing various materials to minimises the consumption of the earth resources. This research examines the implementation of crushed limestone as a coarse aggregate instead of traditional coarse aggregate in Iraq. Besides, rice husk ash was also used in this research project to improve the mechanical and durability properties of the developed concrete using crushed limestone. Two concrete mixtures were developed in this research. The first mixture uses the crushed limestone as aggregated without rice husk ash while the second mixture employs the crushed limestone with rice husk ash. Several tests were conducted to assess the mechanical and durability properties of the developed concrete that are slump and air contents, absorption compressive strength and flexural strength. The outcome highlighted that the concrete developed using crushed limestone has lower mechanical and durability properties compared to normal aggerate (gravel). However, the usage of the rice husk ash improved the properties of the developed concrete using crushed limestone.

Keywords: Limestone, rice husk ash, aggregate, concrete.

1. Introduction

Industrial development and urbanisation have increased the demand for new development and structures. One of the most commonly used structural materials is concrete due to its durability and strength [1, 2]. Concrete is considered an essential element from various types of structures. One of the basic elements of concrete is the aggregate which plays an essential role in the compressive strength of the concrete. The aggregate occupies the largest size of the concrete and the concrete performance is large influenced by the type of aggregate used in it [3]. Yet, this industry of concrete is considered by many researchers as a considerable source of pollution around the world. Concrete produces various types of gases that affect human health and the environment [4-6] such as greenhouse gases which causes global warming. Global warming causes a significant shortage is of freshwater and pollution [7, 8]. Additio nally, concrete production generates huge quantities of polluted water with various types of pollutants disposed of in water bodies[9-12]. These pollutants include suspended solids, organic compounds [13-17]. Accordingly, concrete factory wastewater needs advanced and effective technologies to remove the pollutants from factory effluences like

filtration [18-24], coagulation [25-29], chemical treatment [24, 30-36], and also mix methods [37-43]. In addition to environmental pollution, the large use of concrete in development has led to the extensive use of natural aggregates which led to the depletion of earth resources of natural aggregate. based on the above effects of concrete usage, researchers have been investigating the use of other eco-friendly materials that replace the constituents of concrete to minimises the effects of concrete. Limestone is considered as one of the commonly used material to replace natural aggregates in concrete production [44-46]. The limestone mostly contains calcium carbonate, magnesium carbonate and siliceous materials in its composition [47, 48]. Using limestone as aggregates in concrete significantly minimises the environmental effects of concrete. Besides, the production of limestone is cheaper than the natural aggregates and needs less effort and energy. Besides, the production of limestone aggregate produces significantly lower quantities of pollutants such as carbon dioxide. additionally, limestone aggregate usage in concrete produces more stable concrete and reduces the quantity of concrete waste and enhance the durability and strength of the concrete. This extends the life span of the concrete and reduces the concrete waste [49-52].
The shortage in naturals aggregate has led to a significant increase in limestone usage as aggregate in concrete production around the world [53]. The coarse aggregate largely affects the concrete properties owing to the effects of gradation of the aggregate and the connection between the aggregate and other materials in the concrete [54]. The parameters of the coarse aggregate like size distribution have a significant influence on the strength of produced concrete structures. Researchers showed that the limestone meets the requirements of aggregates and could be adopted to produce concrete mixes. Therefore, the crushed limestone provides a useful alternative to the concrete aggregates that age the gravel and even the sand in concrete mixtures. Researchers [55] examined the usage of the dust of the limestone in concrete. The usage of the dust of the limestone in terms of various properties like compressive strength, absorption, permeability, and others. It was found that fine aggregate can be substituted by limestone dust.
In addition to aggregate replacement, researchers examined the use of other Cementitious materials like silica fume, ground blast furnace slag, fly ash as cement replacement to develop eco-friendly cementitious materials to replace cement in concrete production [45, 56]. The effects of the cementitious materials depend on the dosage of the replacement materials and characteristics. For instance, Abdulredha, Muhsin, Al-Janabi, Alajmi, Gkantou, Amoako-Attah, Al-Jumeily, Mustafina and AlKhayyat [45] used silica fume as cement replacement in concrete production. They showed that the silica fume could be used as cement replacement in concrete development. Shubbar, Jafer, Abdulredha, Al-Khafaji, Nasr, Al Masoodi and Sadique [46] used ground blast furnace slag in concrete production and highlighted that the ground blast furnace slag is a useful replacement for the cement in mortars production. Generally. Researchers reported that the use of cementitious materials as cement replacement generates concrete with a very good characteristic that is comparable to the concrete produced by ordinary cement [57]. This could significantly reduce the negative impact of cement and reduces the depletion of the earth resources.
Based on the above the current research examines the use of crushed limestone and coarse aggregates replacement in concrete production. The use of crushed limestone as coarse aggregate replacement is based on several reasons that are wide availability of the crushed limestone, low production price of the crushed limestone, and low impact of the environmental impact of crashed limestone concrete comparing to ordinary concrete[44-46]. Besides, rice husk ash usage as cement replacement in concrete production was also investigated in concrete incorporating crushed limestone as coarse aggregate.

2. Methodology

2.1. Materials

To reach the goal of this research, several materials were used including ordinary Portland cement, natural coarse aggregates, crushed limestone and rice husk ash. The ordinary Portland cement meets the Iraqi standard No.5:1984 of the Iraqi Organization of Standards for Portland Cement. The cement specific gravity is 3.14. The initial setting times of the cement is 120 minutes while its final meeting time is 250 minutes. Sand is the second component used in cement production which has a specific gravity of 2.62. The particle size distribution of the sand ranged from 200μm to 4.75mm. the size gradation of the sand agrees with the limits recognised by ASTM C33M-18.

In addition, gravel and crushed limestone were used in this research as coarse aggregates. The crushed limestone was obtained from the Al-Noura factory located about 8 km south of the city of Kerbala, Iraq. The physical and chemical characteristics of the gravel and the limestone are presented in tables 1 and 2, respectively. The largest particle size for limestone particles is 25 mm. The standard ASTM C127-12 was adopted to test the characteristics of the gravel and the limestone.

Table 1: Gravel and limestone physical characteristics.

Aggregate type	Bulk Density (g/cm^3)	Specific Gravity	Porosity (%)	Moisture Content (%)
Gravel	1.60	2.70	1.40	0.05
Limestone	1.20	2.55	5.0	0.28

Table 2: Gravel and limestone physical-chemical characteristics.

	CaO	Al_2O_3	MgO	SiO_2HP	Fe_2O_3
limestone	53.728	0.168	1.018	0.814	0.126
Gravel	1.72	6.16	0.06	84.81	1.66

The characteristics of the rice husk ash are presented in table 3. The standard ASTM C618-15 was used to confirm the chemical and physical characteristic of the rice husk ash. From the table, it can be seen the specific gravities of rice husk ash is 2.18.

Table 3: Physical and chemical analysis of the rice husk ash.

Oxide composition	Magnitude
SiO_2 %	86.50 %
Al_2O_3 %	0.15 %
Fe_2O_3 %	0.31 %
SO_3 %	0.14 %
Na_2O %	1.40 %
Strength activity Index	130
Flow table test	100
Specific gravity	2.18

2.2. Testing standards

Several tests were conducted to show the performance of the concrete that contains limestone and rice husk ash. The standard followed to show the characteristic of the concrete include the slump test (ASTM C143), flexural strength test (ASTM C 78-02), compressive strength test (ASTM C39) and absorption test (BS 1881:122-1983).

2.3. Mixtures

Various percentages of the limestone were used to study its effects on the properties of the developed concrete. The limestone replacement ranged from 20% to 80% by weight of the coarse aggregate of the developed concrete. The concrete created from natural gravel was used as a control spaceman for comparison. The reference mixture was designed based on the guidelines of ACI-211.1-91. This is one to make the developed concrete comply with the requirements of workability and compressive strength. The reference mixture contains only natural sand, ordinary Portland cement and gravel. In the first group of developed concrete, the coarse aggregate of gravel is replaced with the crashed limestone with various percentages ranged from 20% to 80%. The mixtures contain crashed limestone were denoted by L followed by the percentage of the coarse limestone used. For example, when 20 of the gravel is replaced with limestone, the mixture is denoted with L20. Four mixtures were developed containing 20%, 40%, 60% and 80% limestone as coarse aggregate denoted by L20, L40, L60 and L80.

Additionally, 15% of the cement was replaced with rice husk ash to show the effects of the added cementitious material on the development of the concrete containing limestone as coarse aggregate. The mixtures contain rice husk ash were denoted with the letter R to highlight the addition of the cementitious material. Based on the above, two groups of mixtures were developed as follow:

- Group one contains various percentages of the limestone only.
- Group two contains various percentages of the limestone in addition to only 20% of the rice husk ash.

The proportion of the mixtures are presented in table 4.

Table 4: mix proportion of the developed mixtures (kg per m^3).

Group	mixtures	Cement	Rice husk ash	Sand	Gravel	Crushed limestone	Water
	L0	450	0	675	1025	--	216
	L20	450	0	675	820	205	225
Group1	L40	450	0	675	615	410	235
	L60	450	0	675	410	615	245
	L80	450	0	675	205	820	250
	L0R	382.5	67.5	675	1025	--	207
	L20R	382.5	67.5	675	820	205	216
Group2	L40R	382.5	67.5	675	615	410	221
	L60R	382.5	67.5	675	410	615	234
	L80R	382.5	67.5	675	205	820	243

3. Results

Several tests were conducted to show the development of the concrete using two replacement materials that are limestone and rice husk ash. Table 5 presents the results of the tests for concrete cubes that uses limestone as aggregates replacement. It is worth mention that the results reported in this table represent the average of three samples. From the table, it can be noticed that the compressive strength of the concrete decrease with an increase in the percentage the limestone replacement. It can be seen that the 7 days compressive strength decreased from 22 MPa to only 14 MPa when the fraction of the limestone increased from 0% to 80% of the coarse aggregate. similarly, the development of the compressive strength after 28 days is comparable to the development of the compressive strength after 7 days. The increases in the quantity of the limestone reduce the compressive strength significantly.

Flexural strength tests result also confirmed that the developed concrete using limestone is weaker compared to normal concrete the employs gravel in its materials. The flexural strength after 28 days of curing decreased from 5.97 MPa to 4.32 MPa when the percentage of the limestone fraction increased from 0% to 80% substitutes. The absorption is also increased with the increase in the percentage of the limestone fraction in the concrete.

Table 5: Tests results for the new concrete mixtures without rising husk ash.

Mixtures		Compressive strength (MPa)		Flexural strength (MPa)		Absorption (%)		Slump (mm)
		7 days	28 days	7 days	28 days	7 days	28 days	
Group 1	L0	22.71	32.21	4.61	5.97	6.15	4.49	100
	L20	18.45	28.40	3.86	5.80	8.46	5.82	92
	L40	15.20	24.01	3.79	5.21	9.79	6.88	87
	L60	15.10	21.85	3.43	4.87	10.07	7.61	85
	L80	14.00	18.97	3.07	4.32	10.41	8.51	77

Table 6 presents the results of the tests for concrete cubes that uses limestone as aggregates replacement and rice husk ash as cement replacement. Similarly, the compressive strength increases from 32.21 MPa (28 days without rice husk ash and limestone) to 34.92MPa (28 days with only 15% rice husk ash). However, the compressive strength of the concrete decrease with an increase in the percentage of the limestone replacement. The 7 days compressive strength decreased from 23.75 MPa to only 14.71 MPa when the fraction of the limestone increased from 0% to 80% of the coarse aggregate. the presence of the rice husk ash increases the compressive strength but the limestone still largely reduces the compressive strength. The flexural strength tests result also is not different from the compressive strength. They developed the concrete using rice husk ash with higher flexural strength. However, the flexural strength is weaker compared to normal concrete the employs gravel in its materials. The flexural strength after 28 days of curing decreased from 7.01 MPa to 5.32 MPa when the percentage of the limestone fraction increased from 0% to 80% substitutes. The absorption is also increased with the increase in the percentage of the limestone fraction in the concrete. However, 15 per cent of the rice husk ash decreased the adsorption percentage by about 8%.

Table 6: Tests results for the new concrete mixtures with rising husk ash.

Mixtures		Compressive strength (MPa)		Flexural strength (MPa)		Absorption (%)		Slump (mm)
		7 days	28 days	7 days	28 days	7 days	28 days	
Group 2	L0R	23.75	34.92	5.14	7.01	6.10	4.12	94
	L20R	19.97	30.11	4.86	6.82	7.34	5.56	90
	L40R	17.45	25.83	4.34	6.11	8.18	6.23	85
	L60R	17.13	24.44	3.91	5.75	8.84	6.75	79
	L80R	14.71	22.76	4.12	5.32	9.64	6.91	74

4. Conclusions

This research is conducted with the aim of investigating the effect of implementing crushed limestone as a coarse aggregate replacement and rice husk ash as cementitious materials. Several tests were conducted to understand the effects of the aforementioned materials on the mechanicals and durability characteristics. These tests include slump test, flexural strength test, compressive strength test and absorption test. The outcome showed that 15% replacement of the rice husk ash with cement increased the compressive strength, flexural strength by about 8% and 15%, respectively. The addition of about 80% of the limestone reduced the compressive strength by about 28%.

References

[1] Kadhim A, Sadique M, Al-Mufti R and Hashim K 2020 Long-term performance of novel high-calcium one-part alkali-activated cement developed from thermally activated lime kiln dust *Journal of Building Engineering* **32** 1-17

[2] Kadhim A, Sadique M, Al-Mufti R and Hashim K 2020 Developing One-Part Alkali-Activated metakaolin/natural pozzolan Binders using Lime Waste as activation Agent *Advances in Cement Research* **32** 1-38

[3] Shubbar A A, Al-Shaer A, AlKizwini R S, Hashim K, Hawesah H A and Sadique M 2019 Investigating the influence of cement replacement by high volume of GGBS and PFA on the mechanical performance of cement mortar. In: *First International Conference on Civil and Environmental Engineering Technologies (ICCEET),* (University of Kufa, Iraq pp 31-8

[4] Shubbar A A, Jafer H, Dulaimi A, Hashim K, Atherton W and Sadique M 2018 The development of a low carbon binder produced from the ternary blending of cement, ground granulated blast furnace slag and high calcium fly ash: An experimental and statistical approach *Construction and Building Materials* **187** 1051-60

[5] Hashim K S, Shaw A, Al Khaddar R, Ortoneda Pedrola M and Phipps D 2017 Defluoridation of drinking water using a new flow column-electrocoagulation reactor (FCER) - Experimental, statistical, and economic approach *Journal of Environmental Management* **197** 80-8

[6] Hashim K S, Idowu I A, Jasim N, Al Khaddar R, Shaw A, Phipps D, Kot P, Pedrola M O, Alattabi A W and Abdulredha M 2018 Removal of phosphate from River water using a new baffle plates electrochemical reactor *MethodsX* **5** 1413-8

[7] Abdulredha M, Abdulridha A, Shubbar A, Alkhaddar R, Kot P and Jordan D 2020 Estimating municipal solid waste generation from service processions during the Ashura religious event. In: *IOP Conference Series: Materials Science and Engineering*: IOP Publishing) p 012075

[8] Aayef A N, Al Masoodi W T M, Kamel R J, Abdulredha M, Almansoori N A, Kot P and Muradov M 2021 An experimental study for adapting electrocoagulation as a technique for fluoride removal from water. In: *IOP Conference Series: Materials Science and Engineering*: IOP Publishing) p 012012

[9] Alyafei A, AlKizwini R S, Hashim K S, Yeboah D, Gkantou M, Al Khaddar R, Al-Faluji D and Zubaidi S L 2020 Treatment of effluents of construction industry using a combined filtration-electrocoagulation method. In: *IOP Conference Series: Materials Science and Engineering*: IOP Publishing) p 012032

[10] Alattabi A W, Harris C, Alkhaddar R, Alzeyadi A and Hashim K 2017 Treatment of Residential Complexes' Wastewater using Environmentally Friendly Technology *Procedia Engineering* **196** 792-9

[11] Alattabi A W, Harris C B, Alkhaddar R M, Hashim K S, Ortoneda-Pedrola M and Phipps D 2017 Improving sludge settleability by introducing an innovative, two-stage settling sequencing batch reactor *Journal of Water Process Engineering* **20** 207-16

[12] Abdulredha M, Al Khaddar R and Jordan D 2017 Hoteliers' attitude towards solid waste source separation through mega festivals: A pilot study in Karbala. In: *International Conference for Doctoral Research*: BUID)

[13] Alenezi A K, Hasan H A, Hashim K S, Amoako-Attah J, Gkantou M, Muradov M, Kot P and Abdulhadi B 2020 Zeolite-assisted electrocoagulation for remediation of phosphate from calcium-phosphate solution. In: *IOP Conference Series: Materials Science and Engineering*: IOP Publishing) p 012031

[14] Alhendal M, Nasir M J, Hashim K S, Amoako-Attah J, Al-Faluji D, Muradov M, Kot P and Abdulhadi B 2020 Cost-effective hybrid filter for remediation of water from fluoride. In: *IOP Conference Series: Materials Science and Engineering*: IOP Publishing) p 012038

[15] Al-Marri S, AlQuzweeni S S, Hashim K S, AlKhaddar R, Kot P, AlKizwini R S, Zubaidi S L and Al-Khafaji Z S 2020 Ultrasonic-Electrocoagulation method for nitrate removal from water. In: *IOP Conference Series: Materials Science and Engineering*: IOP Publishing) p 012073

[16] Grmasha R A, Al-sareji O J, Salman J M, Hashim K S and Jasim I A 2020 Polycyclic Aromatic Hydrocarbons (PAHs) in Urban Street Dust WithinThree Land-Uses of Babylon Governorate, Iraq: Distribution, Sources, andHealth Risk Assessment *Journal of King Saud University - Engineering Sciences* **33**

[17] Abdulredha M, Al Khaddar R, Jordan D, Al-Attabi A and Alzeyadi A 2017 Public participation in solid waste management during mega festivals: A pilot study. In: *WCST World Congress on Sustainable Technologies Proceedings 2017*: Infonomics Society) pp 38-41

[18] Hashim K S, Al-Saati N H, Hussein A H and Al-Saati Z N 2018 An investigation into the level of heavy metals leaching from canal-dreged sediment: a case study metals leaching from dreged sediment. In: *First International Conference on Materials Engineering & Science,* (Istanbul Aydın University (IAU), Turkey pp 12-22

[19] Abdulla G, Kareem M M, Hashim K S, Muradov M, Kot P, Mubarak H A, Abdellatif M and Abdulhadi B 2020 Removal of iron from wastewater using a hybrid filter. In: *IOP Conference Series: Materials Science and Engineering*: IOP Publishing) p 012035

[20] Hashim K S, Khaddar R A, Jasim N, Shaw A, Phipps D, Kota P, Pedrola M O, Alattabi A W, Abdulredha M and Alawsh R 2019 Electrocoagulation as a green technology for phosphate removal from River water *Separation and Purification Technology* **210** 135-44

[21] Hashim K S, AlKhaddar R, Shaw A, Kot P, Al-Jumeily D, Alwash R and Aljefery M H 2020 *Electrocoagulation as an eco-friendly River water treatment method. In Advances in Water Resources Engineering and Management* (Berline: Springer)

[22] Al-Jumeily D, Hashim K, Alkaddar R, Al-Tufaily M and Lunn J 2019 Sustainable and Environmental Friendly Ancient Reed Houses (Inspired by the Past to Motivate the Future). In: *11th International Conference on Developments in eSystems Engineering (DeSE),* (Cambridge, UK pp 214-9

[23] Hashim K S, Ewadh H M, Muhsin A A, Zubaidi S L, Kot P, Muradov M, Aljefery M and Al-Khaddar R 2020 Phosphate removal from water using bottom ash: Adsorption performance, coexisting anions and modelling studies *Water Science and Technology* **83** 1-17

[24] Abdulredha M, al-Khaddar R, Kot P, Jordan D and Abdulridha A 2018 Benchmarking of the Current Solid Waste Management System in Karbala, Iraq, Using Wasteaware Benchmark Indicators. In: *World Environmental and Water Resources Congress 2018: Groundwater, Sustainability, and Hydro-Climate/Climate Change*: American Society of Civil Engineers Reston, VA) pp 40-8

[25] Abdulraheem F S, Al-Khafaji Z S, Hashim K S, Muradov M, Kot P and Shubbar A A 2020 Natural filtration unit for removal of heavy metals from water. In: *IOP Conference Series: Materials Science and Engineering*: IOP Publishing) p 012034

[26] Alenazi M, Hashim K S, Hassan A A, Muradov M, Kot P and Abdulhadi B 2020 Turbidity removal using natural coagulants derived from the seeds of strychnos potatorum: statistical and experimental approach. In: *IOP Conference Series: Materials Science and Engineering*: IOP Publishing) p 012064

[27] Al-Saati N H, Hussein T K, Abbas M H, Hashim K, Al-Saati Z N, Kot P, Sadique M, Aljefery M H and Carnacina I 2019 Statistical modelling of turbidity removal applied to non-toxic natural coagulants in water treatment: a case study *Desalination and Water Treatment* **150** 406-12

[28] Omran I I, Al-Saati N H, Hashim K S, Al-Saati Z N, Patryk K, Khaddar R A, Al-Jumeily D, Shaw A, Ruddock F and Aljefery M 2019 Assessment of heavy metal pollution in the Great Al-Mussaib irrigation channel *Desalination and Water Treatment* **168** 165-74

[29] Abdulredha M, Al Khaddar R, Jordan D, Kot P, Abdulridha A and Hashim K J W M 2018 Estimating solid waste generation by hospitality industry during major festivals: A quantification model based on multiple regression **77** 388-400

[30] Aqeel K, Mubarak H A, Amoako-Attah J, Abdul-Rahaim L A, Al Khaddar R, Abdellatif M, Al-Janabi A and Hashim K S 2020 Electrochemical removal of brilliant green dye from wastewater. In: *IOP Conference Series: Materials Science and Engineering*: IOP Publishing) p 012036

[31] Emamjomeh M M, Mousazadeh M, Mokhtari N, Jamali H A, Makkiabadi M, Naghdali Z, Hashim K S and Ghanbari R 2020 Simultaneous removal of phenol and linear alkylbenzene sulfonate from automotive service station wastewater: Optimization of coupled electrochemical and physical processes *Separation Science and Technology* **55** 3184-94

[32] Hashim K, Kot P, Zubaid S, Alwash R, Al Khaddar R, Shaw A, Al-Jumeily D and Aljefery M 2020 Energy efficient electrocoagulation using baffle-plates electrodes for efficient Escherichia Coli removal from Wastewater *Journal of Water Process Engineering* **33** 101079-86

[33] Mohammed A-H, Hussein A H, Yeboah D, Al Khaddar R, Abdulhadi B, Shubbar A A and Hashim K S 2020 Electrochemical removal of nitrate from wastewater. In: *IOP Conference Series: Materials Science and Engineering*: IOP Publishing) p 012037

[34] Abdulhadi B A, Kot P, Hashim K S, Shaw A and Khaddar R A 2019 Influence of current density and electrodes spacing on reactive red 120 dye removal from dyed water using electrocoagulation/electroflotation (EC/EF) process. In: *First International Conference on Civil and Environmental Engineering Technologies (ICCEET),* (University of Kufa, Iraq pp 12-22

[35] Hashim K S, Al-Saati N H, Alquzweeni S S, Zubaidi S L, Kot P, Kraidi L, Hussein A H, Alkhaddar R, Shaw A and Alwash R 2019 Decolourization of dye solutions by electrocoagulation: an investigation of the effect of operational parameters. In: *First International Conference on Civil and Environmental Engineering Technologies (ICCEET),* (University of Kufa, Iraq pp 25-32

[36] Hashim K S, Hussein A H, Zubaidi S L, Kot P, Kraidi L, Alkhaddar R, Shaw A and Alwash R 2019 Effect of initial pH value on the removal of reactive black dye from water by electrocoagulation (EC) method. In: *2nd International Scientific Conference,* (Al-Qadisiyah University, Iraq pp 12-22

[37] Hashim K S, Ali S S M, AlRifaie J K, Kot P, Shaw A, Al Khaddar R, Idowu I and Gkantou M 2020 Escherichia coli inactivation using a hybrid ultrasonic–electrocoagulation reactor *Chemosphere* **247** 125868-75

[38] Hassan Alnaimi I J I, Abuduljaleel Al-Janabi, Khalid Hashim, Michaela Gkantou, Salah L. Zubaidi, Patryk Kot, Magomed Muradov 2020 Ultrasonic-electrochemical treatment for effluents of concrete plants Ultrasonic-electrochemical treatment for effluents of concrete plants. In: *IOP Conference Series Materials Science and Engineering* (University of Kufa, Najaf, Iraq pp 1-9

[39] Zanki A K, Mohammad F H, Hashim K S, Muradov M, Kot P, Kareem M M and Abdulhadi B 2020 Removal of organic matter from water using ultrasonic-assisted electrocoagulation method. In: *IOP Conference Series: Materials Science and Engineering*: IOP Publishing) p 012033

[40] Abdulhadi B, Kot P, Hashim K, Shaw A, Muradov M and Al-Khaddar R 2021 Continuous-flow electrocoagulation (EC) process for iron removal from water: Experimental, statistical and economic study *Science of The Total Environment* **756** 1-16

[41] Hashim K S, Shaw A, AlKhaddar R, Kot P and Al-Shamma'a3 A 2021 Water purification from metal ions in the presence of organic matter using electromagnetic radiation-assisted treatment *Journal of Cleaner Production* **280**

[42] Idowu I A, Atherton W, Hashim K, Kot P, Alkhaddar R, Alo B I and Shaw A 2019 An analyses of the status of landfill classification systems in developing countries: Sub Saharan Africa landfill experiences *Waste Management* **87** 761-71

[43] Abdulredha M, Kadhim N, Hussein A, Almutairi M, Alkhaddar R, Yeboah D and Hashim K 2021 Zeolite as a natural adsorbent for nitrogenous compounds being removed from water. In: *IOP Conference Series: Materials Science and Engineering*: IOP Publishing) p 012082

[44] Abdulredha M, Kot P, Al Khaddar R, Jordan D, Abdulridha A J E, Development and Sustainability 2020 Investigating municipal solid waste management system performance during the Arba'een event in the city of Kerbala, Iraq **22** 1431-54

[45] Abdulredha M, Muhsin A A, Al-Janabi A, Alajmi B N, Gkantou M, Amoako-Attah J, Al-Jumeily D, Mustafina J and AlKhayyat A 2021 Using SF and CKD as cement replacement materials for producing cement mortar. In: *IOP Conference Series: Materials Science and Engineering*: IOP Publishing) p 012007

[46] Shubbar A A, Jafer H, Abdulredha M, Al-Khafaji Z S, Nasr M S, Al Masoodi Z and Sadique M J J o B E 2020 Properties of cement mortar incorporated high volume fraction of GGBFS and CKD from 1 day to 550 days **30** 101327

[47] O Amin M, AM Khidir M and M Taher A 2014 Manufacturing of Bricks from Soil and Crushed Limestone by Compression-E *J AL-Rafdain Engineering Journal* **22** 24-32

[48] Majdi H S, Shubbar A, Nasr M S, Al-Khafaji Z S, Jafer H, Abdulredha M, Masoodi Z A, Sadique M and Hashim K 2020 Experimental data on compressive strength and ultrasonic pulse velocity properties of sustainable mortar made with high content of GGBFS and CKD combinations *Data in Brief* **31** 105961-72

[49] Shubbar A A, Sadique M, Nasr M S, Al-Khafaji Z S and Hashim K S 2020 The impact of grinding time on properties of cement mortar incorporated high volume waste paper sludge ash *Karbala International Journal of Modern Science* **6** 1-23

[50] Shubbar A A, Sadique M, Shanbara H K and Hashim K 2020 *The Development of a New Low Carbon Binder for Construction as an Alternative to Cement. In Advances in Sustainable Construction Materials and Geotechnical Engineering* (Berlin: Springer)

[51] Abdulredha M, Rafid A, Jordan D and Hashim K 2017 The development of a waste management system in Kerbala during major pilgrimage events: determination of solid waste composition *Procedia Engineering* **196** 779-84

[52] Abdulredha M, Al Khaddar R, Jordan D, Kot P, Abdulridha A and Hashim K 2018 Estimating solid waste generation by hospitality industry during major festivals: A quantification model based on multiple regression *Waste Management* **77** 388-400

[53] Aquino C, Inoue M, Miura H, Mizuta M and Okamoto T 2010 The effects of limestone aggregate on concrete properties *J Construction Building Materials* **24** 2363-8

[54] Kamali-Bernard S, Keinde D and Bernard F 2014 Effect of aggregate type on the concrete matrix/aggregates interface and its influence on the overall mechanical behavior. A numerical study. In: *Key engineering materials*: Trans Tech Publ) pp 14-7

[55] Kürklü G, Görhan G J C and Materials B 2019 Investigation of usability of quarry dust waste in fly ash-based geopolymer adhesive mortar production *J Construction Building Materials* **217** 498-506

[56] Majdi H S, Shubbar A, Nasr M S, Al-Khafaji Z S, Jafer H, Abdulredha M, Al Masoodi Z, Sadique M and Hashim K J D i B 2020 Experimental data on compressive strength and ultrasonic pulse velocity properties of sustainable mortar made with high content of GGBFS and CKD combinations

[57] Wang D, Ju Y, Shen H and Xu L 2019 Mechanical properties of high performance concrete reinforced with basalt fiber and polypropylene fiber *J Construction Building Materials* **197** 464-73

Removal of culturable bacteria from municipal wastewater using electrocoagulation

M Abdulredha [1], N R Kadhim[2], M A A Naji[3,*]

[1] Department of Civil Engineering, College of Engineering, University of Kerbala, Kerbala, Iraq.
[2] Department of Civil Engineering, College of Engineering, University of Warith AL-Anbiya'a, Kerbala, Iraq.
[3] State Company for Oil Projects, Ministry of Oil, Baghdad, Iraq
Email: muhammed.r@uokerbala.edu.iq

Abstract. Many countries around the world are faced with a serious event that is the pollution of water and wastewater with biological pollutions. This is particularly true in developing nations because of the weak treatment and disposal practice of wastewater. The weak economic situation and low financial support in many low-income countries significantly reduce the development and the uses of innovative water and wastewater treatment methods. This has led to increases and enlarges the issue of water pollution biological pollutants. The goal of this research is to implement the electrocoagulation technique as a cheap and cost-effective technology to eliminate or reduce the per cent of biological pollutant from wastewater. The coagulants are produced in situ by passing a current between two electrodes. A sample of the wastewater was collected from the Kerbala sewage treatment plant, which is located in the southeast of Kerbala city, Iraq. Iron electrodes were used to produce the coagulants in this research study. Besides, the impact of several variables on the performance of the electrocoagulation unit was investigated including the spacing between electrodes and current density. The results revealed that the *E. coli* bacteria as biological indicator were destroyed after 40 minutes of treatment using iron electrodes with a spacing of 5mm and a current density of 2 mA/cm^2. Besides, the outcome showed that an initial pH level of 6.0 is suitable for biological pollutants treatment using electrocoagulation.

Keywords: Electrocoagulation, E. coli, wastewater.

1. Introduction

There is an agreement among researchers around the world that the earth is currently faced with serious and grave water scarcity [1, 2]. Researchers attributed the shortage in water resources to many factors. This includes the limited quantity of fresh water on the earth surfaces which is accessible by a human. researchers reported that only 1% of the water found in the plant each is fresh and consumable by human and other livings[2-4]. Besides, the disposal of large quantities of wastewater to surface water bodies, particularly freshwater bodies largely reduced the quality and quantity of freshwater around the world [5-7]. Additionally, the change in the plant earth climate is also reported as important aspects that affect the quantity of freshwater around the world [8, 9]. For example, due to the drought which happens in many parts of the world, the consumption rate of freshwater is significantly increased[2, 5, 10-13].

Besides, climate change influences the natural distribution of water around the world and increases the deficiency of potable water[14-16].

The disposal of wastewater such as industrial and municipal wastewater largely reduced the presence of freshwater around the world [8, 9, 12]. The disposal of wastewater to water bodies is considered as one of the main concerns to water researchers due to the huge quantities of wastewater disposal to water bodies and the diverse types of pollutants found in the wastewater which ends up in freshwater stream [17, 18]. For example, water bodies in many parts around China are significantly polluted with pesticides as more than 300 million tons of pesticides are used in agricultural activities per year which ends up in water bodies [19-22]. In addition to pesticides, many other pollutants can find their way to water bodies like heavy metals, dyes, and organic pollutants [23-26]. The concentration of such pollutants largely increased in water bodies over the last few decades owing to industrialisation and urbanization [10, 27-30]. Researchers showed that the concentration of phosphorus pollutants in river and other water bodies has raised by more than 70% over the last two decades [17, 31-33]. Biological pollutants are considered the main freshwater pollutants and have significant attention among researchers because of their effect on health and the environment [34-38].

The focus on biological contaminants has intensely raised over the last decades [2, 18, 37]. Since the 1850s, scholars start to explore and expand our understanding of the occurrence and the spread of various type of pathogens, particularly through water bodies. Researchers focused on dangerous diseases like gastrointestinal disorders and their spread through water[4, 6]. Despite these efforts to understand the spread of biological pollution, biological pollution is still noticeably increasing because of many aspects like urbanization. Researchers found the more than 30 wells distributed in a rural area in Bangladesh were contaminated with *E. coli* [10, 12, 30]. This problem is amplified by the lack or the weakness of the advanced treatment method, particularly in poor and developing countries [34, 39]. Accordingly, many pieces of research have been conducted to reduce biological contamination using cheap and effective treatment technologies like chlorination and coagulation [1, 39-42]. Nevertheless, the chemical methods generate various types of by-products that have a toxic impact and reduces their usages in wastewater treatment.

This research is developed with the goal of using an iron-based electrocoagulation unit to remove or eliminate the biological pollution of wastewater. The influences of electrical current density and electrodes spacing on the performance of the iron-based electrocoagulation unit have been investigated to achieve higher removal efficiency of *E. coli.* The low cost, simple operation, possible development and low sludge production were the main aspects that motivate the author to select the electrocoagulation method for biological treatment[16, 21]. This significantly reduces the need for sludge treatment and final disposal in landfill sites [43, 44].

2. Methodology

2.1. Iron-based electrocoagulation unit

The coagulation unit used in this research is developed for several materials. The electrocoagulation is a rectangular vessel that has 120 mm length, 84 mm depth and 114 mm long and built from transparent plastic. The size and the dimension of the electrocoagulation reactor are based on previous research studies that employed the electrocoagulation method for wastewater treatment [4, 33]. The achieve proper water mixing and improved water treatment 8 iron electrodes. The iron electrodes were laid in a way that enables the water to be mixed while flow through the unit. Each used electrode has a dimension of 96 mm length and 72 mm depth and placed vertically in the electrocoagulation unit. Iron has been selected to be used as electrodes materials in this research because it is cheaply available around the world compared to other materials [45-47]. Besides, the dimension and the placement of the electrodes in the reactor were adopted based on previous research studies [4].

2.2. Wastewater sample

Actual wastewater samples were used in this research to achieve the goal of this research project of disinfecting the wastewater using the electrocoagulation method. The wastewater samples were obtained from the Kerbala sewage treatment plant, which is located in the southeast of Kerbala city, Iraq. The obtained sables were concentrated and diluted with denoised water before conducting the tests.
To test the performance of the electrocoagulation method in terms of wastewater disinfection, the E. coli bacteria were used as the indicator. The American Public Health Association method was adopted to conduct the planting and the growth of the E. coli bacteria. The diluted water was planted to determine the number of E.coli colonies and compared with the number of E.coli colonies found in treated wastewater.

2.3. *Wastewater disinfection calculations*
To calculate the removal efficiency of the electrocoagulation unit, it is very important the find the original number of the *E. coli* colonies in the wastewater. In this research, the original number of the *E. coli* colonies in the wastewater is denoted with A1. Then, 600 mL of the wastewater was treated using iron electrodes in an electrocoagulation reactor. The current density and the spacing of the electrodes were changed to achieve the highest disinfection performance. The current densities used in this research 1, 2, and 3 mA/cm^2 and the spacing of the electrodes ranged from 5 to 10 mm. It is worth mentioning that the pH level of the wastewater samples was kept at 6 as the normal pH level of the municipal wastewater is 6 [6, 33, 38]. After each disinfection trial, a sample obtained from the disinfected wastewater were planted and incubated to calculate the residual number of the *E. coli* colonies (A2). The following formula was used to calculate the removal efficiency of the electrocoagulation unit:

$$R\% = \frac{A1-A2}{A2} \times 100 \quad (1)$$

3. Results

3.1. *Effects of current density change*
Many researchers showed that the current density is considered a very important element in water and wastewater treatment [9, 38]. The current density controls the generation of the oxides from the electrodes which significantly affects the removal efficiency of the contaminants from the wastewater. Besides, the current density also controls the generation of the bubbles that help to float contaminants to the surface of the wastewater [4, 48]. Thus, the effects of the current density on the removal efficiency of the biological pollutants from actual wastewater have been examined. The current density was changed from 1 to 3 mA/cm^2 and the other variables were maintained constants. The pH, treatment time, and electrodes spacing were maintained at 6, 40 minutes, and 5 mm, respectively. Figure 1 presents the results of the effect of changing the current density on the removal efficiency of the biological pollutants from wastewater. The removal efficiency of the *E. coli* raised from 75% at 1 mA/cm^2 to 100% at 3 mA/cm^2. This can be attributed to the increase in the production of oxides and the generation bobbles with the increased current density. However, the increase in the current density largely affects the treatment cost making the treatment method uneconomical.

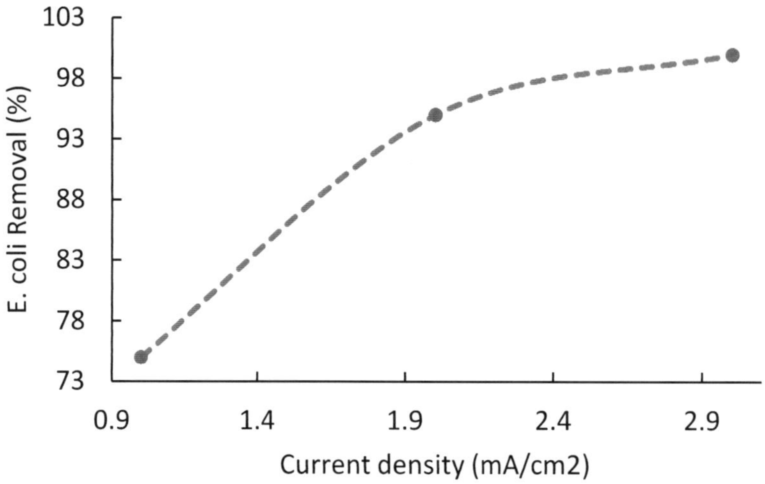

Figure 1: The effects of current density on the removal efficiency of the biological pollutants from wastewater.

Figure 2 presents the effects of the current density of the power usage. The figure showed that the increases in the current density significantly increases the power usage which reduces the economic integrity of the electrocoagulation unit using iron electrodes. The power usages significantly raised from 0.8 kW.h/m^3 to 7.3 kW.h/m^3 when the current density increased from 1 to 3 mA/cm^2. Based on the above, it can be said that a current density of 2 mA/cm^2 provides high removal and significantly lower cost comparing to 3 mA/cm^2.

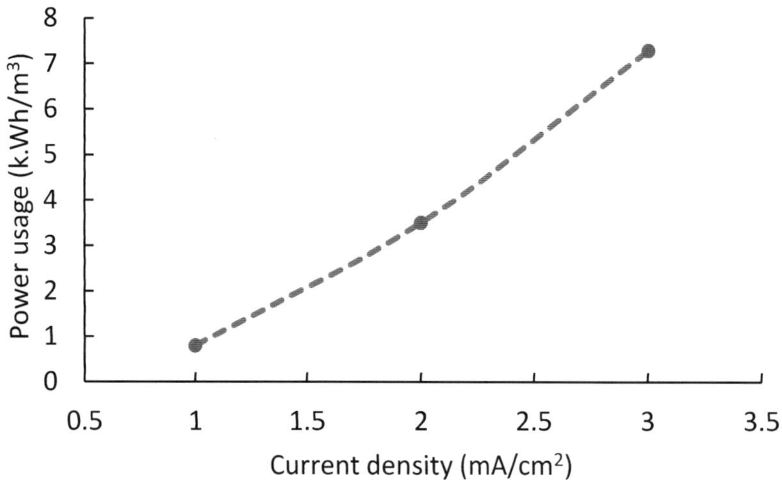

Figure 2: The effects of the current density of power usage.

3.2. Effects of spacing on the removal efficiency

Studies on electrocoagulation usage in water and wastewater treatment showed that the spacing between electrodes has a large effect on the performance of the method [4, 28]. The spacing between electrodes affects the electrical resistance as smaller spacings have smaller electrical resistance [34, 48].

Accordingly, this could affect the removal efficiency of the electrocoagulation unit. based on the above, the influence of the spacing between the electrodes has been examined in this research. The experimental parameters that are current density, treatment time and pH level were kept constants at 2 mA/cm^2, 40 and 6.0, respectively. The spacing between the electrodes was changed from 5 to 10 mm.
Figure 3 presents the effect of spacing on the removal efficiency of *E. coli* from the wastewater. It can be seen the larger spacing between the electrodes the lower removal efficiency. From the figure, we can see that the removal efficiency increased from 89% to 99% when the spacing decreased from 10 to 5 mm. The electrical resistance increased with the increase in the spacing between the electrodes which leads to a change in the removal efficiency of the *E. coli* [5, 14, 47].

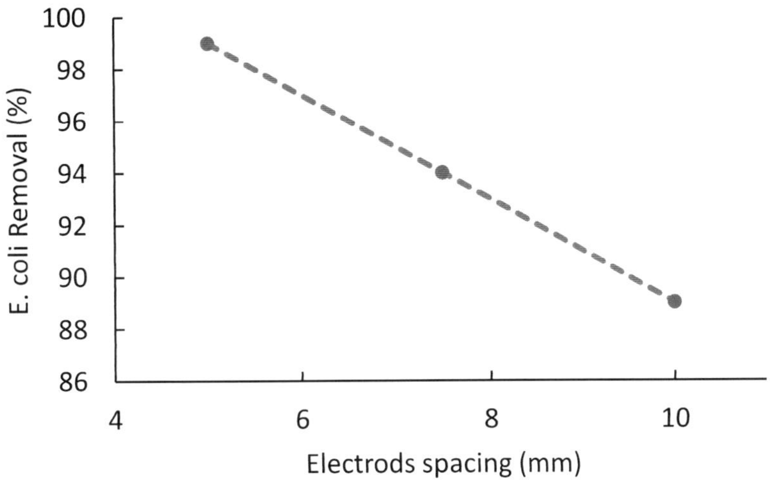

Figure 3: The effect of electrode spacing on the removal efficiency of the biological pollutants from wastewater.

4. Conclusion

This study examines the application of the electrocoagulation method as a disinfectant for wastewater treatment plant effluent using iron electrodes with different spacing and current density. The outcome of this research revealed that the iron can be used as an electrode in the electrocoagulation method to eliminate the presence of culturable bacteria in wastewater treatment plant effluent. The required time to remove the culturable bacteria from the wastewater effluent was 40 minutes. Based on the results, it can be concluded that the current density and the spacing between electrodes influence the removal efficiency of the culturable bacteria in wastewater treatment plant effluent.

References

[1] Abdulredha M, Al Khaddar R and Jordan D 2017 Hoteliers' attitude towards solid waste source separation through mega festivals: A pilot study in Karbala. In: *International Conference for Doctoral Research*: BUID)

[2] Al-Sareji O J, Abdulredha M, Mubarak H A, Grmasha R A, Alnowaishry A, Kot P, Al-Khaddar R and AlKhayyat A 2021 Copper removal from water using carbonized sawdust. In: *IOP Conference Series: Materials Science and Engineering*: IOP Publishing) p 012015

[3] Abdulredha M, Al Khaddar R, Jordan D, Al-Attabi A and Alzeyadi A 2017 Public participation in solid waste management during mega festivals: A pilot study. In: *WCST World Congress on Sustainable Technologies Proceedings 2017*: Infonomics Society) pp 38-41
[4] Jawad S F, Saddam N S, Adaami Q J, Kareem M M, Abdulredha M, Mubarak H A, Kot P, Gkantou M and AlKhayyat A 2021 Dye removal from textile wastewater using solar-powered electrocoagulation reactor. In: *IOP Conference Series: Materials Science and Engineering*: IOP Publishing) p 012016
[5] Abdulredha M, Rafid A, Jordan D and Hashim K J P E 2017 The development of a waste management system in Kerbala during major pilgrimage events: determination of solid waste composition **196** 779-84
[6] Alattabi A W, Harris C, Alkhaddar R, Alzeyadi A and Abdulredha M J P e 2017 Online monitoring of a sequencing batch reactor treating domestic wastewater **196** 800-7
[7] Abdulredha M, Muhsin A A, Al-Janabi A, Alajmi B N, Gkantou M, Amoako-Attah J, Al-Jumeily D, Mustafina J and AlKhayyat A 2021 Using SF and CKD as cement replacement materials for producing cement mortar. In: *IOP Conference Series: Materials Science and Engineering*: IOP Publishing) p 012007
[8] Aayef A N, Al Masoodi W T M, Kamel R J, Abdulredha M, Almansoori N A, Kot P and Muradov M 2021 An experimental study for adapting electrocoagulation as a technique for fluoride removal from water. In: *IOP Conference Series: Materials Science and Engineering*: IOP Publishing) p 012012
[9] Abdulredha M, Kadhim N, Hussein A, Almutairi M, Alkhaddar R, Yeboah D and Hashim K 2021 Zeolite as a natural adsorbent for nitrogenous compounds being removed from water. In: *IOP Conference Series: Materials Science and Engineering*: IOP Publishing) p 012082
[10] Abdulredha M, Kot P, Al Khaddar R, Jordan D, Abdulridha A J E, Development and Sustainability 2020 Investigating municipal solid waste management system performance during the Arba'een event in the city of Kerbala, Iraq **22** 1431-54
[11] Farhan S L, Jasim I A and Al-Mamoori S K 2019 The transformation of the city of Najaf, Iraq: Analysis, reality and future prospects *Journal of Urban Regeneration & Renewal* **13** 160-71
[12] Isra'a S S, Al-Janabi A, Abdulredha M, Alkandari A, Abdellatif M and Yeboah D 2021 Reusing of furnace bottom ash as an adsorbent for phosphate removal from water. In: *IOP Conference Series: Materials Science and Engineering*: IOP Publishing) p 012006
[13] Shubbar A A, Jafer H, Abdulredha M, Al-Khafaji Z S, Nasr M S, Al Masoodi Z and Sadique M J J o B E 2020 Properties of cement mortar incorporated high volume fraction of GGBFS and CKD from 1 day to 550 days **30** 101327
[14] Al-Marri S, AlQuzweeni S S, Hashim K S, AlKhaddar R, Kot P, AlKizwini R S, Zubaidi S L and Al-Khafaji Z S 2020 Ultrasonic-Electrocoagulation method for nitrate removal from water. In: *IOP Conference Series: Materials Science and Engineering*: IOP Publishing) p 012073
[15] Alyafei A, AlKizwini R S, Hashim K S, Yeboah D, Gkantou M, Al Khaddar R, Al-Faluji D and Zubaidi S L 2020 Treatment of effluents of construction industry using a combined filtration-electrocoagulation method. In: *IOP Conference Series: Materials Science and Engineering*: IOP Publishing) p 012032
[16] Aqeel K, Mubarak H A, Amoako-Attah J, Abdul-Rahaim L A, Al Khaddar R, Abdellatif M, Al-Janabi A and Hashim K S 2020 Electrochemical removal of brilliant green dye from wastewater. In: *IOP Conference Series: Materials Science and Engineering*: IOP Publishing) p 012036
[17] Abdulredha M, Abdulridha A, Shubbar A, Alkhaddar R, Kot P and Jordan D 2020 Estimating municipal solid waste generation from service processions during the Ashura religious event. In: *IOP Conference Series: Materials Science and Engineering*: IOP Publishing) p 012075
[18] Abdulredha M, Al Khaddar R, Jordan D, Kot P, Abdulridha A and Hashim K J W M 2018 Estimating solid waste generation by hospitality industry during major festivals: A quantification model based on multiple regression **77** 388-400

[19] Zhang Y, Han S, Liang D, Shi X, Wang F, Liu W, Zhang L, Chen L, Gu Y and Tian Y 2014 Prenatal exposure to organophosphate pesticides and neurobehavioral development of neonates: a birth cohort study in Shenyang, China *Egyptian journal of petroleum* **27** 1275-90.

[20] Abdulraheem F S, Al-Khafaji Z S, Hashim K S, Muradov M, Kot P and Shubbar A A 2020 Natural filtration unit for removal of heavy metals from water. In: *IOP Conference Series: Materials Science and Engineering*: IOP Publishing) p 012034

[21] Emamjomeh M M, Mousazadeh M, Mokhtari N, Jamali H A, Makkiabadi M, Naghdali Z, Hashim K S and Ghanbari R 2020 Simultaneous removal of phenol and linear alkylbenzene sulfonate from automotive service station wastewater: Optimization of coupled electrochemical and physical processes *Separation Science and Technology* **55** 3184-94

[22] Majdi H S, Shubbar A, Nasr M S, Al-Khafaji Z S, Jafer H, Abdulredha M, Al Masoodi Z, Sadique M and Hashim K J D i B 2020 Experimental data on compressive strength and ultrasonic pulse velocity properties of sustainable mortar made with high content of GGBFS and CKD combinations

[23] Hashim K S, Shaw A, Al Khaddar R, Pedrola M O and Phipps D 2017 Energy efficient electrocoagulation using a new flow column reactor to remove nitrate from drinking water - Experimental, statistical, and economic approach *Journal of Environmental Management* **196** 224-33

[24] Hashim K S, Shaw A, Al Khaddar R, Ortoneda Pedrola M and Phipps D 2017 Defluoridation of drinking water using a new flow column-electrocoagulation reactor (FCER) - Experimental, statistical, and economic approach *Journal of Environmental Management* **197** 80-8

[25] Alattabi A W, Harris C, Alkhaddar R, Alzeyadi A and Hashim K 2017 Treatment of Residential Complexes' Wastewater using Environmentally Friendly Technology *Procedia Engineering* **196** 792-9

[26] Alattabi A W, Harris C B, Alkhaddar R M, Hashim K S, Ortoneda-Pedrola M and Phipps D 2017 Improving sludge settleability by introducing an innovative, two-stage settling sequencing batch reactor *Journal of Water Process Engineering* **20** 207-16

[27] Alenazi M, Hashim K S, Hassan A A, Muradov M, Kot P and Abdulhadi B 2020 Turbidity removal using natural coagulants derived from the seeds of strychnos potatorum: statistical and experimental approach. In: *IOP Conference Series: Materials Science and Engineering*: IOP Publishing) p 012064

[28] Hashim K S, AlKhaddar R, Shaw A, Kot P, Al-Jumeily D, Alwash R and Aljefery M H 2020 *Electrocoagulation as an eco-friendly River water treatment method. In Advances in Water Resources Engineering and Management* (Berline: Springer)

[29] Hassan Alnaimi I J I, Abuduljaleel Al-Janabi, Khalid Hashim, Michaela Gkantou, Salah L. Zubaidi, Patryk Kot, Magomed Muradov 2020 Ultrasonic-electrochemical treatment for effluents of concrete plants Ultrasonic-electrochemical treatment for effluents of concrete plants. In: *IOP Conference Series Materials Science and Engineering* (University of Kufa, Najaf, Iraq pp 1-9

[30] ALWAN H H, SALEH L A, AL-MOHAMMED F M, ABDULREDHA M A J J o E S and Technology 2020 EXPERIMENTAL PREDICTION OF THE DISCHARGE COEFFICIENTS FOR RECTANGULAR WEIR WITH BOTTOM ORIFICES **15** 3265-80

[31] Hashim K S, Idowu I A, Jasim N, Al Khaddar R, Shaw A, Phipps D, Kot P, Pedrola M O, Alattabi A W and Abdulredha M 2018 Removal of phosphate from River water using a new baffle plates electrochemical reactor *MethodsX* **5** 1413-8

[32] Hashim K S, Khaddar R A, Jasim N, Shaw A, Phipps D, Kot P, Pedrola M O, Alattabi A W, Abdulredha M and Alawsh R 2019 Electrocoagulation as a green technology for phosphate removal from River water *Separation and Purification Technology* **210** 135-44

[33] Hashim K S, Al Khaddar R, Jasim N, Shaw A, Phipps D, Kot P, Pedrola M O, Alattabi A W, Abdulredha M, Alawsh R J S and Technology P 2019 Electrocoagulation as a green technology for phosphate removal from River water **210** 135-44

[34] Hashim K, Kot P, Zubaid S, Alwash R, Al Khaddar R, Shaw A, Al-Jumeily D and Aljefery M 2020 Energy efficient electrocoagulation using baffle-plates electrodes for efficient Escherichia Coli removal from Wastewater *Journal of Water Process Engineering* **33** 101079-86
[35] Hashim K S, Ali S S M, AlRifaie J K, Kot P, Shaw A, Al Khaddar R, Idowu I and Gkantou M 2020 Escherichia coli inactivation using a hybrid ultrasonic–electrocoagulation reactor *Chemosphere* **247** 125868-75
[36] Al-Jumeily D, Hashim K, Alkaddar R, Al-Tufaily M and Lunn J 2019 Sustainable and Environmental Friendly Ancient Reed Houses (Inspired by the Past to Motivate the Future). In: *11th International Conference on Developments in eSystems Engineering (DeSE),* (Cambridge, UK pp 214-9
[37] Abdulredha M, al-Khaddar R, Kot P, Jordan D and Abdulridha A 2018 Benchmarking of the Current Solid Waste Management System in Karbala, Iraq, Using Wasteaware Benchmark Indicators. In: *World Environmental and Water Resources Congress 2018: Groundwater, Sustainability, and Hydro-Climate/Climate Change*: American Society of Civil Engineers Reston, VA) pp 40-8
[38] Hashim K S, Idowu I A, Jasim N, Al Khaddar R, Shaw A, Phipps D, Kot P, Pedrola M O, Alattabi A W and Abdulredha M J M 2018 Removal of phosphate from River water using a new baffle plates electrochemical reactor **5** 1413-8
[39] Hashim K S, Shaw A, AlKhaddar R, Kot P and Al-Shamma'a3 A 2021 Water purification from metal ions in the presence of organic matter using electromagnetic radiation-assisted treatment *Journal of Cleaner Production* **280**
[40] Ghernaout D, Touahmia M and Aichouni M 2019 Disinfecting water: Electrocoagulation as an efficient process *Applied Engineering* **3** 1-12
[41] Hashim K S, Ewadh H M, Muhsin A A, Zubaidi S L, Kot P, Muradov M, Aljefery M and Al-Khaddar R 2020 Phosphate removal from water using bottom ash: Adsorption performance, coexisting anions and modelling studies *Water Science and Technology* **83** 1-17
[42] Al-Anbari R, Alnakeeb A, Abdulredha M J E and Journal T 2013 Landfill site selection for Kerbala municipal solid wastes by using geographical information system techniques **32** 13
[43] Idowu I A, Atherton W, Hashim K, Kot P, Alkhaddar R, Alo B I and Shaw A 2019 An analyses of the status of landfill classification systems in developing countries: Sub Saharan Africa landfill experiences *Waste Management* **87** 761-71
[44] Majdi H S, Shubbar A, Nasr M S, Al-Khafaji Z S, Jafer H, Abdulredha M, Masoodi Z A, Sadique M and Hashim K 2020 Experimental data on compressive strength and ultrasonic pulse velocity properties of sustainable mortar made with high content of GGBFS and CKD combinations *Data in Brief* **31** 105961-72
[45] Hashim K S, Hussein A H, Zubaidi S L, Kot P, Kraidi L, Alkhaddar R, Shaw A and Alwash R 2019 Effect of initial pH value on the removal of reactive black dye from water by electrocoagulation (EC) method. In: *2nd International Scientific Conference,* (Al-Qadisiyah University, Iraq pp 12-22
[46] Zanki A K, Mohammad F H, Hashim K S, Muradov M, Kot P, Kareem M M and Abdulhadi B 2020 Removal of organic matter from water using ultrasonic-assisted electrocoagulation method. In: *IOP Conference Series: Materials Science and Engineering*: IOP Publishing) p 012033
[47] Abdulhadi B A, Kot P, Hashim K S, Shaw A and Khaddar R A 2019 Influence of current density and electrodes spacing on reactive red 120 dye removal from dyed water using electrocoagulation/electroflotation (EC/EF) process. In: *First International Conference on Civil and Environmental Engineering Technologies (ICCEET),* (University of Kufa, Iraq pp 12-22
[48] Hashim K S, Shaw A, Al Khaddar R, Pedrola M O and Phipps D 2017 Iron removal, energy consumption and operating cost of electrocoagulation of drinking water using a new flow column reactor *Journal of Environmental Management* **189** 98-108

Replacement of cement with industrial by-products in cement mortar: An experimental investigation

A. Aljobbory[1,*] **and H. Al-Lammi**[2]
[1] Ministry of Construction, Baghdad, Iraq
Email: Aljobory_81@yahoo.com

Abstract. The manufacture of cement has been described as a significant exporter of carbon gases pollution and other contaminants, with (1000) kilograms of Cement reportedly producing one ton of Carbon gases. The effects of these gases in the atmosphere on public health and the atmosphere are negative impacts. As a result, industrials by-products become more eco-friend and sustainable comparing with Portland cement in terms of cement use and emissions of carbon gases. The effect of partial substitution in mortar by ground granulated blast furnace slag (GGBS) and coal fly ash (CFA) is the subject of this investigation. three separate GGBS and CFA blends have been utilized, and they have been checked at 7 days, 14 days, and 28 days curing age. to investigate the mechanical properties of the three mixtures, the compression test has been utilized. the experiment indicates that raising the percentage of GGBS and CFA in the mixture reduced the compressive strength of these mixes. 20 to 33% was the reduction in the compressive strength of the mixtures with supplementary Cementous materials, which might be considered sufficient.

1. Introduction

Nowadays, have a significant role in building manufacturing, and the essential part of it is Portland cement [1-3]. However, the manufacture of Portland cement has been an increasing issue as a significant volume of carbon gases, and some other pollutants are emitted into the environment throughout this phase [4-7]. The available literature, reveals that the production of cement produces around 6% to 8% of the overall Co2 in the environment [8-10]. To put it another way, generating (1000) kilogram of Portland cement is expected to yield one ton of Carbon dioxide. Carbon dioxide and other pollutants gases caused a significant increase in air temperature and the atmosphere overall. This last phenomenon endangers global safety by increasing water contamination and causing drought in certain countries of the planet [11-13]. Furthermore, concrete factories generate vast volumes of waste effluents with high suspended particle concentrations, pH values (close to 11), and sedimentation [14-17], both of which pose a hazard to marine organisms in aquatic ecosystems [16, 18, 19]. As a result of the unfavorable effects of the construction materials factories on global safety and the atmosphere, effective management strategies, as well as articulate wastewater technologies such as filtering [20-27], combination technologies [24, 28-30], flocculation [13, 16, 23, 31-40], and electro-chemical techniques [26, 41-48], as well as drainage alternatives [49-51], are needed.

Several researchers have currently been tasked with discovering substitute supplementary cementitious materials that can supplement a portion of Portland cement, such as ground granulated blast furnace slab (GGBS) and coal fly ash (CFA) [52]. Utilizing these admixtures as a partial substitute material could save power, money and reduce greenhouse gas emissions. Because of the pozzolanic characteristic of coal fly ash (CFA) and ground granulated blast furnace slab (GGBS), they are widely utilizing to partially replace Portland cement. As a result, utilizing these ingredients has a substantial impact on the cement mixture characteristics [52, 53].

The influences of ground granulated blast furnace slab (GGBS) and coal fly ash (CFA) on conventional cement were investigated in this research. The major aim of this report is to check how these compounds influence the mechanical properties of cement mixtures at 7 days, 14 days, and 28 days of curing period.

2. Experimental program

To compare the compression strength of mixtures produced by partially substituting Portland cement with crushed fly ash (CFA) and ground granulated blast furnace slab (GGBS), several experiments have been carried out. The ratios of crushed fly ash (CFA) and ground granulated blast furnace slab (GGBS) utilized in samples, as well as the design of mixtures, are subsequently Neglected.

2.1 Materials

The ground granulated blast furnace slab (GGBS) and coal fly ash (CFA) have been utilized as cementitious substitution substances in this research.
The ground granulated blast furnace slab (GGBS) is a secondary product produced during the production of ferrous metals, and it is characterized as a "nonmetallic material composed mainly of Ca2SiO4 and other compounds that is formed in a molten state concurrently with iron in a kiln," according to ACI-116R. Silicates comprise the essential component of GGBS. Slag cooling is specifically responsible for producing multiple kinds of blast furnace slag required by various end-users. Slag's characteristics change dramatically as it cools. GGBS is a non-toxic substance that can be used to make high-value, user-friendly cementitious materials for a variety of structural engineering fields [54, 55].

CFA is a secondary product of crushed burning coal in thermal energy stations. It is collected as quite thin, mainly circular glassy grains from the evaporator coils by dust collecting devices till they are emitted into the air from the energy stations. the fly ash atoms have a different dimension, where their radius varies from less than 0.5μm to more than 75μm. It has a smoother texture than conventional cement. The forms and proportional quantities of incombustible content in the coal utilized decide the chemical structure of fly ash [56, 57].

Portland cement was used as the basic binding material in this experiment because of its excellent mechanical properties, which aid in its binding with all other components of the mixture. BS-EN 196-2:2013 has been utilized to Evaluation the cement used in this project. Figures 1 and 2 list the chemical characteristics of cementitious admixtures. Coal fly ash, Portland Cement, and the ground granulated blast furnace meet the requirements of BS-EN 450-1:2012, BS-EN 197-1:2011, and BS-EN 15167-1:2006.

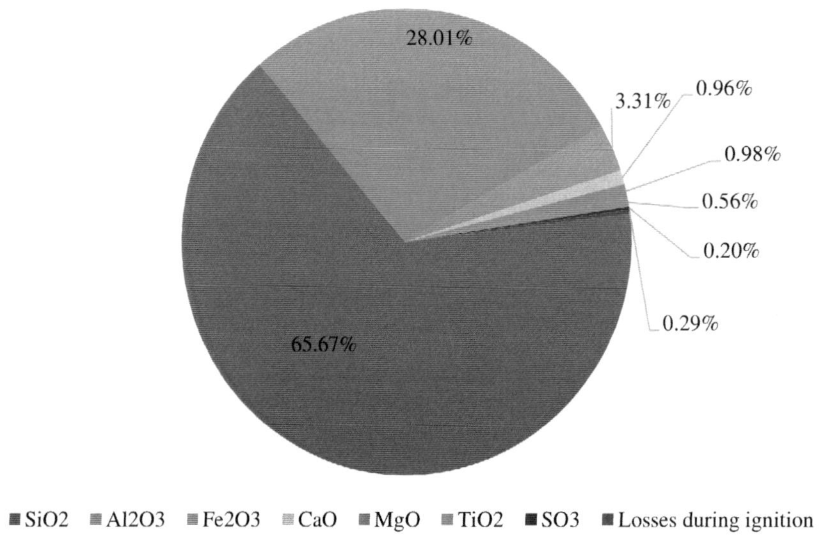

Figure 1: The chemical composition of the CFA

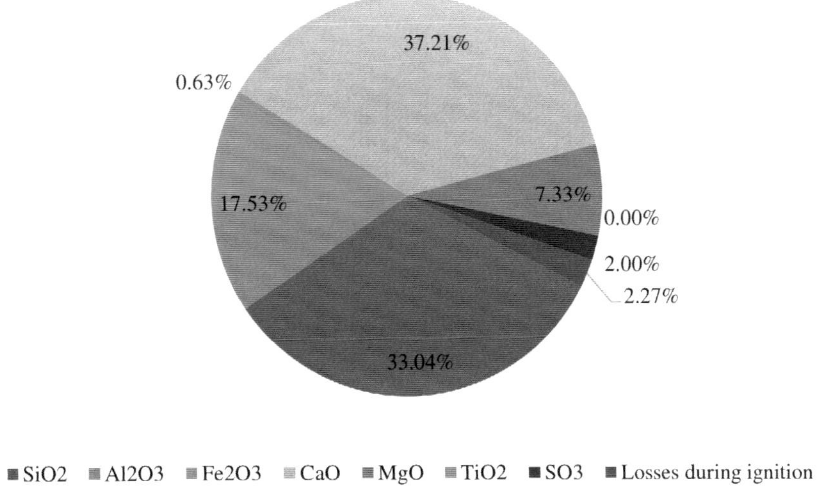

Figure 2: The chemical composition of the GGBS

2.2 Methods of testing

Three specimens (100x100x100 mm) were Pour and measured for each sample to evaluate the compressive strength of Portland cement mixtures. Checking the cubes just with Portland cement and after that with separate substitution ratios of ground granulated blast furnace slab and coal fly ash to understand its effects on the compressive strength of the cement mixes. after pouring the cement cubes and handled them with care, the cubes have been molded and curing it's at different periods of 7,14and 28 days. then, the compressive strength was calculated for each cube.

2.3 Design of Mixtures

The quantity of aggregate, supplementary cementitious products, water, and Portland cement, in a mortar mixture, is determined using the mix design approach used in this study. To match the classification scales, natural sand was selected. for all of the mixes, two parts of Portland cement to five parts of fine aggregate was the ratio of sand to binding materials (CFA + cement + GGBS). As well as, the percentage of water to binding materials was 0.4. Figure 3 presents the value of each element of mixing as well as the blend style proportions. To assess the strength of composite mortar, all cubes were subjected to a compression examination. Ground granulated blast furnace slab and coal fly ash could be utilized to evaluate the mechanical characteristics of cement mortar by 10x10x10 cm cubes. After 7, 14, and 28 days of cure, these cubes were examined by a compression instrument. Cubes were evaluated after gradually applying the load from the compression machine till these cubes were a failure. The impacted force was divided by an area of (10000) square millimeters to measure the compressive resistance of these cubes

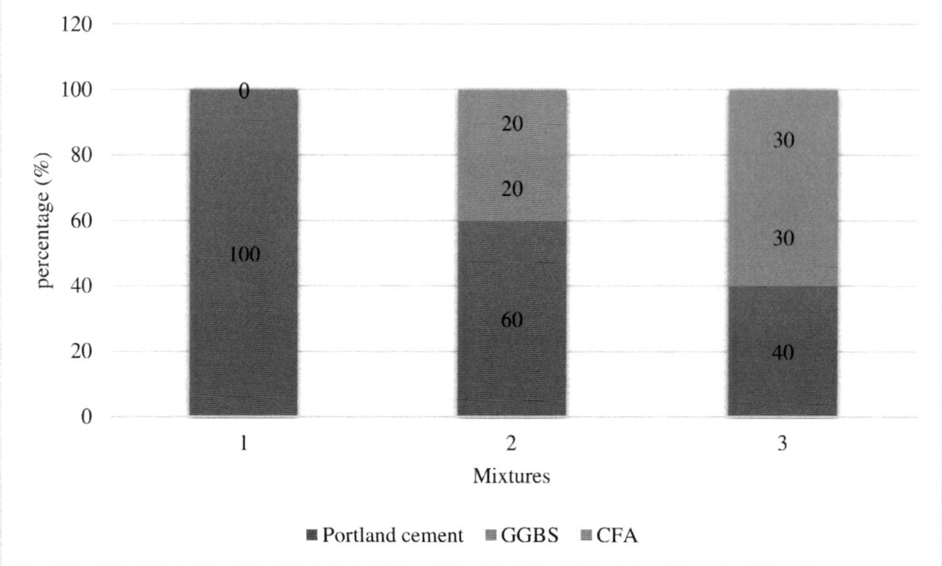

Figure 3:The design mixtures developed in this research.

3. Results

Table (1) and Figure 4 show the examining results of the cubes for cement mortar with partial substitution of ground granulated blast furnace slab and coal fly ash at varying proportion and curing times, after examining them with a compression machine.

Table 1: the compressive strengths of the developed mortars at tested ages.

Test No.	Portland cement %	CFA %	GGBS %	Average of compressive strength (MPa) at		
				7 days	14 days	28 days
1	100	0	0	9.9	11	13.5
2	60	20	20	7.24	9.07	10.4
3	40	30	30	5.30	7.25	7.77

It is shown that partially adding ground granulated blast furnace slab and coal fly ash in cementitious mixtures leads to a decrease in the compressive strengths of these mixtures. When comparing mixture No. (2) with the third mixture, it is evident that mixture (2) has a lower compressive intensity decrease. For such as, after 7 days of pretreatment, the compressive intensity of the blends was reduced by 26% and 46%, respectively, as compared to the first blend (Figure 4). On another side, utilizing 40% partially replacing content reduced the compressive efficiency of cementitious mortar by 46% after 7 days, but this percentage slipped to 30% after 4 weeks. This suggests that it was at a young age, ground granulated blast furnace slab and coal fly ash are inert materials that require duration to interact with the elements of a cement mortar. To summarize, the limited substitution of CFA and GGBS can be achieved at a rate of 20 to 33% to achieve better quality while reducing emissions. This is because supplementary cementitious materials usually minimize the compressive properties of cementitious mixtures, which is an essential component in the manufacturing gel (C-S-H) in cement mixes. The findings show that specimens that have been cured for 28 days have a greater compressive characteristic. To explanation, as mention above the curing age affects the compressive strength of cementitious mortars. where, curing period influences the benefit and improvement of C-S-H, which contributes to a reduction in the number of Internal spaces or permeability in the mortar matrix, which influences the properties of concrete and increases the strength development of these mixes. When comparing mixture three at 14 and 28 days to the same specimens at 7 days, the improvement in compressive intensity was 27% and 32%, consecutive.

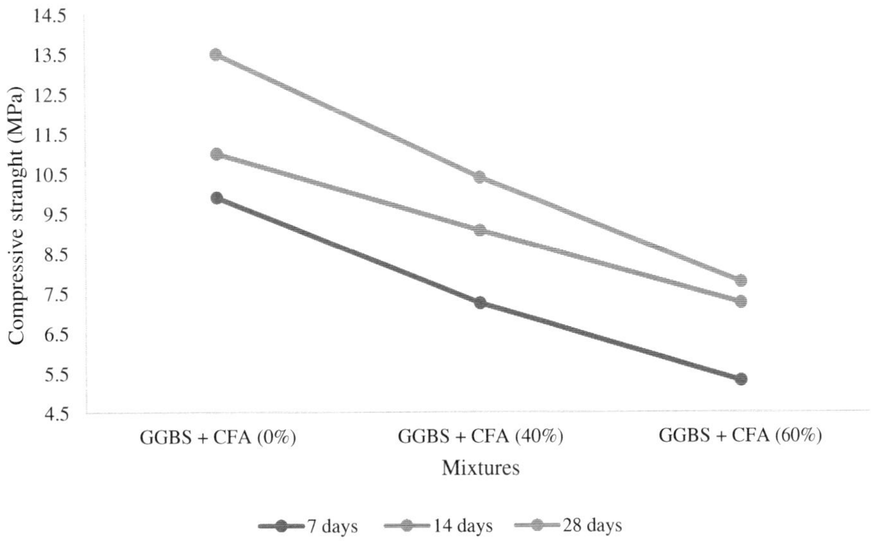

Figure 4: Shows the compressive strength of blends after 7, 14, and 28 days of curing.

The current study looked at the compressive strength of concrete specimens for 28 days; however, further research is needed to look at extended periods and equate the findings to conventional concrete samples of the same age. The built-in sensors can also be utilized for that [58, 59], as sensors have demonstrated high-quality results in a variety of engineering sector, including communications [60] and water quality [61].

4. Conclusions

Drawn from the results of this research, it can be concluded that when ground granulated blast furnace slab and coal fly ash are utilized to substitute Portland cement in mortars, the compressive intensity reduces as the ratio of GGBS and CFA is increased. also, Whenever the ratio of the cementitious material maximized, the curing time should be longer to ensure achieve better compression characteristics. Utilizing about 20 to 33% as a partial substitute for cement may be suitable values, as increasing this value reduces the concrete's compressive characteristics.

Since the current report just looked at the compressive strength of concrete samples for 28 days, further ages must be studied in the future. The built-in sensors can be utilized for that as sensors have demonstrated a high-quality result in a variety of engineering sector.

References

[1] Shubbar A A, Jafer H, Dulaimi A, Hashim K, Atherton W and Sadique M 2018 The development of a low carbon binder produced from the ternary blending of cement, ground granulated blast furnace slag and high calcium fly ash: An experimental and statistical approach *Construction and Building Materials* **187** 1051-60

[2] Al-Jumeily D, Hashim K, Alkaddar R, Al-Tufaily M and Lunn J 2019 Sustainable and Environmental Friendly Ancient Reed Houses (Inspired by the Past to Motivate the Future). In: *11th International Conference on Developments in eSystems Engineering (DeSE),* (Cambridge, UK pp 214-9

[3] Al-Anbari R, Alnakeeb A, Abdulredha M J E and Journal T 2013 Landfill site selection for Kerbala municipal solid wastes by using geographical information system techniques **32** 13

[4] Kadhim A, Sadique M, Al-Mufti R and Hashim K 2020 Developing One-Part Alkali-Activated metakaolin/natural pozzolan Binders using Lime Waste as activation Agent *Advances in Cement Research* **32** 1-38

[5] Shubbar A A, Sadique M, Nasr M S, Al-Khafaji Z S and Hashim K S 2020 The impact of grinding time on properties of cement mortar incorporated high volume waste paper sludge ash *Karbala International Journal of Modern Science* **6** 1-23

[6] Grmasha R A, Al-sareji O J, Salman J M, Hashim K S and Jasim I A 2020 Polycyclic Aromatic Hydrocarbons (PAHs) in Urban Street Dust WithinThree Land-Uses of Babylon Governorate,

Iraq: Distribution, Sources, andHealth Risk Assessment *Journal of King Saud University - Engineering Sciences* **33** 1-18

[7] Abdulredha M, Al Khaddar R and Jordan D 2017 Hoteliers' attitude towards solid waste source separation through mega festivals: A pilot study in Karbala. In: *International Conference for Doctoral Research*: BUID)

[8] Majdi H S, Shubbar A, Nasr M S, Al-Khafaji Z S, Jafer H, Abdulredha M, Masoodi Z A, Sadique M and Hashim K 2020 Experimental data on compressive strength and ultrasonic pulse velocity properties of sustainable mortar made with high content of GGBFS and CKD combinations *Data in Brief* **31** 105961-72

[9] Shubbar A A, Sadique M, Shanbara H K and Hashim K 2020 *The Development of a New Low Carbon Binder for Construction as an Alternative to Cement. In Advances in Sustainable Construction Materials and Geotechnical Engineering* (Berlin: Springer)

[10] Abdulredha M, Al Khaddar R, Jordan D, Al-Attabi A and Alzeyadi A 2017 Public participation in solid waste management during mega festivals: A pilot study. In: *WCST World Congress on Sustainable Technologies Proceedings 2017*: Infonomics Society) pp 38-41

[11] Abdulredha M, Al Khaddar R, Jordan D, Kot P, Abdulridha A and Hashim K J W M 2018 Estimating solid waste generation by hospitality industry during major festivals: A quantification model based on multiple regression **77** 388-400

[12] Hashim K S, Idowu I A, Jasim N, Al Khaddar R, Shaw A, Phipps D, Kot P, Pedrola M O, Alattabi A W and Abdulredha M J M 2018 Removal of phosphate from River water using a new baffle plates electrochemical reactor **5** 1413-8

[13] Hashim K S, Al Khaddar R, Jasim N, Shaw A, Phipps D, Kot P, Pedrola M O, Alattabi A W, Abdulredha M, Alawsh R J S and Technology P 2019 Electrocoagulation as a green technology for phosphate removal from River water **210** 135-44

[14] Alyafei A, AlKizwini R S, Hashim K S, Yeboah D, Gkantou M, Al Khaddar R, Al-Faluji D and Zubaidi S L 2020 Treatment of effluents of construction industry using a combined filtration-electrocoagulation method. In: *IOP Conference Series: Materials Science and Engineering*: IOP Publishing) p 012032

[15] Hassan Alnaimi I J I, Abuduljaleel Al-Janabi, Khalid Hashim, Michaela Gkantou, Salah L. Zubaidi, Patryk Kot, Magomed Muradov 2020 Ultrasonic-electrochemical treatment for effluents of concrete plants Ultrasonic-electrochemical treatment for effluents of concrete plants. In: *IOP Conference Series Materials Science and Engineering* (University of Kufa, Najaf, Iraq pp 1-9

[16] Alhendal M, Nasir M J, Hashim K S, Amoako-Attah J, Al-Faluji D, Muradov M, Kot P and Abdulhadi B 2020 Cost-effective hybrid filter for remediation of water from fluoride. In: *IOP Conference Series: Materials Science and Engineering*: IOP Publishing) p 012038

[17] Abdulredha M, Rafid A, Jordan D and Hashim K J P E 2017 The development of a waste management system in Kerbala during major pilgrimage events: determination of solid waste composition **196** 779-84

[18] Alattabi A W, Harris C, Alkhaddar R, Alzeyadi A and Hashim K 2017 Treatment of Residential Complexes' Wastewater using Environmentally Friendly Technology *Procedia Engineering* **196** 792-9

[19] Alattabi A W, Harris C B, Alkhaddar R M, Hashim K S, Ortoneda-Pedrola M and Phipps D 2017 Improving sludge settleability by introducing an innovative, two-stage settling sequencing batch reactor *Journal of Water Process Engineering* **20** 207-16

[20] Abdulraheem F S, Al-Khafaji Z S, Hashim K S, Muradov M, Kot P and Shubbar A A 2020 Natural filtration unit for removal of heavy metals from water. In: *IOP Conference Series: Materials Science and Engineering*: IOP Publishing) p 012034

[21] Alenazi M, Hashim K S, Hassan A A, Muradov M, Kot P and Abdulhadi B 2020 Turbidity removal using natural coagulants derived from the seeds of strychnos potatorum: statistical and experimental approach. In: *IOP Conference Series: Materials Science and Engineering*: IOP Publishing) p 012064

[22] Alenezi A K, Hasan H A, Hashim K S, Amoako-Attah J, Gkantou M, Muradov M, Kot P and Abdulhadi B 2020 Zeolite-assisted electrocoagulation for remediation of phosphate from calcium-phosphate solution. In: *IOP Conference Series: Materials Science and Engineering*: IOP Publishing) p 012031

[23] Hashim K, Kot P, Zubaid S, Alwash R, Al Khaddar R, Shaw A, Al-Jumeily D and Aljefery M 2020 Energy efficient electrocoagulation using baffle-plates electrodes for efficient Escherichia Coli removal from Wastewater *Journal of Water Process Engineering* **33** 101079-86

[24] Hashim K S, Al-Saati N H, Hussein A H and Al-Saati Z N 2018 An investigation into the level of heavy metals leaching from canal-dreged sediment: a case study metals leaching from dreged sediment. In: *First International Conference on Materials Engineering & Science,* (Istanbul Aydın University (IAU), Turkey pp 12-22

[25] Shubbar A A, Jafer H, Abdulredha M, Al-Khafaji Z S, Nasr M S, Al Masoodi Z and Sadique M J J o B E 2020 Properties of cement mortar incorporated high volume fraction of GGBFS and CKD from 1 day to 550 days **30** 101327

[26] Aayef A N, Al Masoodi W T M, Kamel R J, Abdulredha M, Almansoori N A, Kot P and Muradov M 2021 An experimental study for adapting electrocoagulation as a technique for fluoride removal from water. In: *IOP Conference Series: Materials Science and Engineering*: IOP Publishing) p 012012

[27] Al-Sareji O J, Abdulredha M, Mubarak H A, Grmasha R A, Alnowaishry A, Kot P, Al-Khaddar R and AlKhayyat A 2021 Copper removal from water using carbonized sawdust. In: *IOP Conference Series: Materials Science and Engineering*: IOP Publishing) p 012015

[28] Hashim K S, Shaw A, Al Khaddar R, Ortoneda Pedrola M and Phipps D 2017 Defluoridation of drinking water using a new flow column-electrocoagulation reactor (FCER) - Experimental, statistical, and economic approach *Journal of Environmental Management* **197** 80-8

[29] Hashim K S, Shaw A, Al Khaddar R, Pedrola M O and Phipps D 2017 Iron removal, energy consumption and operating cost of electrocoagulation of drinking water using a new flow column reactor *Journal of Environmental Management* **189** 98-108

[30] Hashim K S, Shaw A, Al Khaddar R, Pedrola M O and Phipps D 2017 Energy efficient electrocoagulation using a new flow column reactor to remove nitrate from drinking water - Experimental, statistical, and economic approach *Journal of Environmental Management* **196** 224-33

[31] Abdulla G, Kareem M M, Hashim K S, Muradov M, Kot P, Mubarak H A, Abdellatif M and Abdulhadi B 2020 Removal of iron from wastewater using a hybrid filter. In: *IOP Conference Series: Materials Science and Engineering*: IOP Publishing) p 012035

[32] Al-Marri S, AlQuzweeni S S, Hashim K S, AlKhaddar R, Kot P, AlKizwini R S, Zubaidi S L and Al-Khafaji Z S 2020 Ultrasonic-Electrocoagulation method for nitrate removal from water. In: *IOP Conference Series: Materials Science and Engineering*: IOP Publishing) p 012073

[33] Abdulhadi B A, Kot P, Hashim K S, Shaw A and Khaddar R A 2019 Influence of current density and electrodes spacing on reactive red 120 dye removal from dyed water using electrocoagulation/electroflotation (EC/EF) process. In: *First International Conference on Civil and Environmental Engineering Technologies (ICCEET),* (University of Kufa, Iraq pp 12-22

[34] Hashim K S, Al-Saati N H, Alquzweeni S S, Zubaidi S L, Kot P, Kraidi L, Hussein A H, Alkhaddar R, Shaw A and Alwash R 2019 Decolourization of dye solutions by electrocoagulation: an investigation of the effect of operational parameters. In: *First International Conference on Civil and Environmental Engineering Technologies (ICCEET),* (University of Kufa, Iraq pp 25-32

[35] Hashim K S, Hussein A H, Zubaidi S L, Kot P, Kraidi L, Alkhaddar R, Shaw A and Alwash R 2019 Effect of initial pH value on the removal of reactive black dye from water by electrocoagulation (EC) method. In: *2nd International Scientific Conference,* (Al-Qadisiyah University, Iraq pp 12-22

[36] Al-Saati N H, Hussein T K, Abbas M H, Hashim K, Al-Saati Z N, Kot P, Sadique M, Aljefery M H and Carnacina I 2019 Statistical modelling of turbidity removal applied to non-toxic natural coagulants in water treatment: a case study *Desalination and Water Treatment* **150** 406-12

[37] Omran I I, Al-Saati N H, Hashim K S, Al-Saati Z N, Patryk K, Khaddar R A, Al-Jumeily D, Shaw A, Ruddock F and Aljefery M 2019 Assessment of heavy metal pollution in the Great Al-Mussaib irrigation channel *Desalination and Water Treatment* **168** 165-74

[38] Abdulredha M, Muhsin A A, Al-Janabi A, Alajmi B N, Gkantou M, Amoako-Attah J, Al-Jumeily D, Mustafina J and AlKhayyat A 2021 Using SF and CKD as cement replacement materials for producing cement mortar. In: *IOP Conference Series: Materials Science and Engineering*: IOP Publishing) p 012007

[39] Isra'a S S, Al-Janabi A, Abdulredha M, Alkandari A, Abdellatif M and Yeboah D 2021 Reusing of furnace bottom ash as an adsorbent for phosphate removal from water. In: *IOP Conference Series: Materials Science and Engineering*: IOP Publishing) p 012006

[40] Jawad S F, Saddam N S, Adaami Q J, Kareem M M, Abdulredha M, Mubarak H A, Kot P, Gkantou M and AlKhayyat A 2021 Dye removal from textile wastewater using solar-powered electrocoagulation reactor. In: *IOP Conference Series: Materials Science and Engineering*: IOP Publishing) p 012016

[41] Aqeel K, Mubarak H A, Amoako-Attah J, Abdul-Rahaim L A, Al Khaddar R, Abdellatif M, Al-Janabi A and Hashim K S 2020 Electrochemical removal of brilliant green dye from wastewater. In: *IOP Conference Series: Materials Science and Engineering*: IOP Publishing) p 012036

[42] Mohammed A-H, Hussein A H, Yeboah D, Al Khaddar R, Abdulhadi B, Shubbar A A and Hashim K S 2020 Electrochemical removal of nitrate from wastewater. In: *IOP Conference Series: Materials Science and Engineering*: IOP Publishing) p 012037

[43] Abdulhadi B, Kot P, Hashim K, Shaw A, Muradov M and Al-Khaddar R 2021 Continuous-flow electrocoagulation (EC) process for iron removal from water: Experimental, statistical and economic study *Science of The Total Environment* **756** 1-16

[44] Hashim K S, Shaw A, AlKhaddar R, Kot P and Al-Shamma'a A 2021 Water purification from metal ions in the presence of organic matter using electromagnetic radiation-assisted treatment *Journal of Cleaner Production* **280**

[45] Hashim K S, AlKhaddar R, Shaw A, Kot P, Al-Jumeily D, Alwash R and Aljefery M H 2020 *Electrocoagulation as an eco-friendly River water treatment method. In Advances in Water Resources Engineering and Management* (Berline: Springer)

[46] Emamjomeh M M, Mousazadeh M, Mokhtari N, Jamali H A, Makkiabadi M, Naghdali Z, Hashim K S and Ghanbari R 2020 Simultaneous removal of phenol and linear alkylbenzene sulfonate from automotive service station wastewater: Optimization of coupled electrochemical and physical processes *Separation Science and Technology* **55** 3184-94

[47] Abdulredha M, Kot P, Al Khaddar R, Jordan D, Abdulridha A J E, Development and Sustainability 2020 Investigating municipal solid waste management system performance during the Arba'een event in the city of Kerbala, Iraq **22** 1431-54

[48] Majdi H S, Shubbar A, Nasr M S, Al-Khafaji Z S, Jafer H, Abdulredha M, Al Masoodi Z, Sadique M and Hashim K J D i B 2020 Experimental data on compressive strength and ultrasonic pulse velocity properties of sustainable mortar made with high content of GGBFS and CKD combinations

[49] Abdulredha M, Rafid A, Jordan D and Hashim K 2017 The development of a waste management system in Kerbala during major pilgrimage events: determination of solid waste composition *Procedia Engineering* **196** 779-84

[50] Idowu I A, Atherton W, Hashim K, Kot P, Alkhaddar R, Alo B I and Shaw A 2019 An analyses of the status of landfill classification systems in developing countries: Sub Saharan Africa landfill experiences *Waste Management* **87** 761-71

[51] Abdulredha M, Al Khaddar R, Jordan D, Kot P, Abdulridha A and Hashim K 2018 Estimating solid waste generation by hospitality industry during major festivals: A quantification model based on multiple regression *Waste Management* **77** 388-400

[52] Shubbar A A, Al-Shaer A, AlKizwini R S, Hashim K, Hawesah H A and Sadique M 2019 Investigating the influence of cement replacement by high volume of GGBS and PFA on the mechanical performance of cement mortar. In: *First International Conference on Civil and Environmental Engineering Technologies (ICCEET),* (University of Kufa, Iraq pp 31-8

[53] Kadhim A, Sadique M, Al-Mufti R and Hashim K 2020 Long-term performance of novel high-calcium one-part alkali-activated cement developed from thermally activated lime kiln dust *Journal of Building Engineering* **32** 1-17

[54] Shumuye E and Jun Z 2018 A review on Ground Granulated Blast Slag GGBS in Concrete. In: *Eighth International Conference On Advances in Civil and Structural Engineering-CSE,*

[55] Suresh D, Nagaraju K J I j o m and engineering c 2015 Ground granulated blast slag (GGBS) in concrete–a review **12** 76-82

[56] Taoufiq L, Laamyem A, Essediqi E, Monkade M and Zradba A J J M E S 2018 Recycling coal fly ash and coal bottom ash from Moroccan thermal power plant in concrete manufacturing **9** 1312-7

[57] Nagrockienė D, Rutkauskas A J C and Materials B 2019 The effect of fly ash additive on the resistance of concrete to alkali silica reaction **201** 599-609

[58] Gkantou M, Muradov M, Kamaris G S, Hashim K, Atherton W and Kot P 2019 Novel Electromagnetic Sensors Embedded in Reinforced Concrete Beams for Crack Detection *Sensors* **19** 5175-89

[59] Teng K H, Kot P, Muradov M, Shaw A, Hashim K, Gkantou M and Al-Shamma'a A 2019 Embedded Smart Antenna for Non-Destructive Testing and Evaluation (NDT&E) of Moisture Content and Deterioration in Concrete *Sensors* **19** 547-59

[60] Ryecroft S, Shaw A, Fergus P, Kot P, Hashim K, Moody A and Conway L 2019 A First Implementation of Underwater Communications in Raw Water Using the 433 MHz Frequency Combined with a Bowtie Antenna *Sensors* **19** 1813-23

[61] Ryecroft S P, shaw A, Fergus P, Kot P, Hashim K and Conway L 2019 A Novel Gesomin Detection Method Based on Microwave Spectroscopy. In: *12th International Conference on Developments in eSystems Engineering (DeSE),* (Kazan, Russia pp 429-33

IOP Conf. Series: Materials Science and Engineering 1184 (2021) 012009 doi:10.1088/1757-899X/1184/1/012009

Petroleum effluent treatment using electrocoagulation: A case study of phenols removal

N R Kadhim[1,*], **M Abdulredha** [2,*]

[1] Department of Civil Engineering, College of Engineering, University of Warith AL-Anbiya'a, Kerbala, Iraq.
[2] Department of Civil Engineering, College of Engineering, University of Kerbala, Kerbala, Iraq.

Email: M-abdulrazak@yahoo.com

Abstract. The petroleum industry is considered as one of the main sources of water pollution. Large amounts of refinery wastewater polluted with various types of pollutants are disposed of into water bodies causing significant negative effects on the environment. Therefore, researchers investigated the application of many treatment methods and technologies to reduce the effects of refinery wastewater. This research project aims to employ the electrocoagulation technique to remove the phenols pollutants from refinery wastewater using aluminium electrodes. Besides, effects of experimental paraments that are electricals current density, the spacing between the electrodes used, and the treatment time on the removal of the phenols have been examined in this research. batch flow experiments were used to remove the phenols from the wastewater. Results showed that phenols concentration in petroleum refinery wastewater can be reduced using the electrocoagulation method. The electrocoagulation unit that using aluminium as electrodes was able to reduce the concentration of the phenols by about 52%. The best removal efficiency was achieved after 120 treatment time using a current density of 6 mA/cm^2 and electrodes spacing of 20 mm. However, the effects of other experiment parameters such as the initial concentration of the phenols need to be investigated.

Keywords: Petroleum refinery wastewater, phenols, electrocoagulation.

1. Introduction

The petroleum industry is one of the main and most important industries around the world since it reparents the main source of energy in many developed and developing countries around the world [1-3]. Besides, the petroleum industry represents the main source of income for many countries around the world [4, 5]. Nevertheless, this industry is considered as one of the main sources of pollution which have a large impact on the environment. This industry consumes large amounts of water and produces huge quantises of wastewater from various units including desalination and cooling units [6-8]. It was shown that by researchers that about 300L of water are needed to produce 1 barrel of crude oil. Diya'uddeen, Daud and Aziz [9] showed that over the current decade more than 33 million barrel of wastewater is produced per day by the petroleum industry around the world. Besides, the current energy usage trends indicate that the pollution generated from the petroleum industry is growing [2, 9, 10].

The pollutants found in the refinery wastewater largely depends on the origins (the refinery process that uses water). The wastewater could contain various types of pollutants such as organic materials, phenols, hydrocarbons, heavy metals, greases, radioactive pollutants, sulphate, cyanides etc. [3-5, 11, 12]. The phenolic complexes are considered very common pollutants which can be found almost in all types of wastewater produced from refinery wastewater. Researchers reported that the concentration of the phenolic compounds in refinery wastewater could reach up to 200 mg/L [13-15]. As the allowable concentration limit for phenol is 1 mg/L, the high concentration of the phenols in the refinery wastewater could cause a severe impact on the environment, ecosystem, and human health [16-18]. Also, the soils around the refineries normally are heavily polluted and the soils need complex treatment processes that are very expensive [8, 19]. The effects of refinery wastewater pollution on human health include serious illnesses or poisoning effects [20-22]. It is very urgent to treat the huge amounts of refinery wastewater to meet the increasing demand for freshwater supply due to climate change. Research on refinery wastewater treatment significantly increased due to the changes in the climate find an appropriate technology to remove pollutants from refinery wastewaters. Refinery wastewater treatment methods ranged from simple to very advanced treatment methods that combine physical, chemical, biological technologies to remove the pollutants from the refinery wastewater [23, 24]. These technologies include coagulation-flocculation, photocatalytic-oxidation, biodegradation, adsorption, and membrane which were used to deduce the impact of one or multiple pollutants that can be found in refinery wastewater [25, 26]. The pollutants removed using these technologies include phosphates [27-29], nitrates [25, 26], microorganisms [30, 31] organic materials [32-34], and heavy metals [35-38]. Nevertheless, many of these techniques are either expensive or ineffective to completely remove or meet the disposal standards limits because of the change in the concentration or the composition of the pollutants in the wastewater produced from the petroleum industry.

Additionally, a growing number of researchers are interested in the electrocoagulation method and proved that this method can remove a wide range of pollutants from various types of wastewater including heavy metals, phosphates, organic compounds, colours etc. [13, 16]. Researchers showed that the electrocoagulation technique is cheap, easy to operate, does not consume chemicals, and require small space [6, 27]. Besides, the sludge of the electrocoagulation method contains many heavy metals making the sludge suitable for recycling, particularly in civil engineering applications [39-41]. This could significantly reduce the amount of waste transfer to landfills and preserver landfill areas for other types of wastes [10, 42-49].

This research is conducted to examine the applicability of using the electrocoagulation method in removing the phenols from petroleum refinery wastewater effluent. The selection of the electrocoagulation technique is based on its attractiveness and advantages listed earlier such as needless to chemical, cheap operating etc. Besides, phenol is considered in this research as it is considered one of the most common pollutants found in petroleum refinery wastewater effluent.

2. Material and methods

Wastewater effluent from a petroleum refinery factory was treated using an electrocoagulation reactor (ER). The ER is rectangular with a dimension of 9 cm in width, 15 cm in-depth, and 25 cm in length containing four aluminium electrodes. Each electrode is perforated aluminium plates with dimensions of15 cm in height and 9 cm in width containing 42 holes of 2.5 mm radius. The arrangement of the electrodes and in the rector and holes in each electrode mixes the wastewater during treatment processes and enhances the removal efficiency.

The wastewater used in this research to evaluate the performance of the ER contains 120 mg/l of phenols. The electrical current density (ECD) as well as the aluminium electrodes spacing (AES) were changed to investigates the effects of these two experimental parameters on the performance of the ER in terms of phenols removal from petroleum refinery wastewater. The properties of the wastewater used in this research are presented in table 1.

Table 1. The properties of the used petroleum refinery wastewater samples.

Parameter	Value	Unit
Colour	Brown	---
Conductivity	9.64	ms/cm
pH	6.7	--
Phenols	120	Mg/l
Salinity	0.14	%
TSS	115	Mg/l

Three desire ECDs that are 4 mA/cm^2, 6 mA/cm^2, and 8 mA/cm^2 were supplied ER to remove the phenols from the refinery wastewater by implementing a DC rectifier. Besides, three distances between the electrodes were used in this research are 20mm, 25mm, and 30mm to examine the effect of the AES on the remove performance of the ER. The distances were changed manually.

The removal process of the phenols from petroleum refinery wastewater was performed by pumping the wastewater in a continuous manner through the ER and the aluminium electrodes. The time used to treat the phenols from the wastewater was ranged from 1 to 120 minutes. During that time, samples of 15 mL of the were collected from the treated wastewater at 5 minutes to investigate the removal performed of the ER. The remaining phenols in the wastewater were determined using a spectrophotometer (Hach Lange DR 2800) and the standard cuvette test LCK 346 or LCK 345 which were selected based on the residual concentration of phenols. The temperature of the wastewater used in this research were kept equal to room temperature of 20 ±1^{0}C. Based on the above methods the removal efficiency of the phenols from refinery wastewater and be calculated by implementing equation 1. The I_0 and I represent the initial and final concentration of the phenols in the refinery wastewater in mg/L, respectively.

$$\text{Removal Efficiency\%} = \frac{I_0 - I}{I_0} \times 100\% \qquad (1)$$

3. Results and discussion

3.1. Influence of ECD

The density of the current in the electrocoagulation method is considered a very important aspect as it is affecting the rate of the coagulants production which significantly affects the removal performance of the ER [3, 50, 51]. Based on this, the effect of the ECD effect of the removal performance of the phenol from refinery wastewater has been investigated in this research. This has been done by treating wastewater samples containing 120 ppm of the pollutant using various ECDs. The ECDs used in this research were 4 mA/cm^2, 6 mA/cm^2, and 8 mA/cm^2. Besides, the AES were kept at 20 mm and the treatment time were 45 minutes. The results (figure 1) highlighted that the increase in the ECD increases the removal efficiency of the phenols from the refinery wastewater. The removal efficiency of the phenols increased from 26% to more than 40% when the current density increased from 4 to 8 mA/cm after 45 minutes of treatment. This increase in the removal efficiency of the phenol can be regraded to the increase in the rete of coagulant generation.

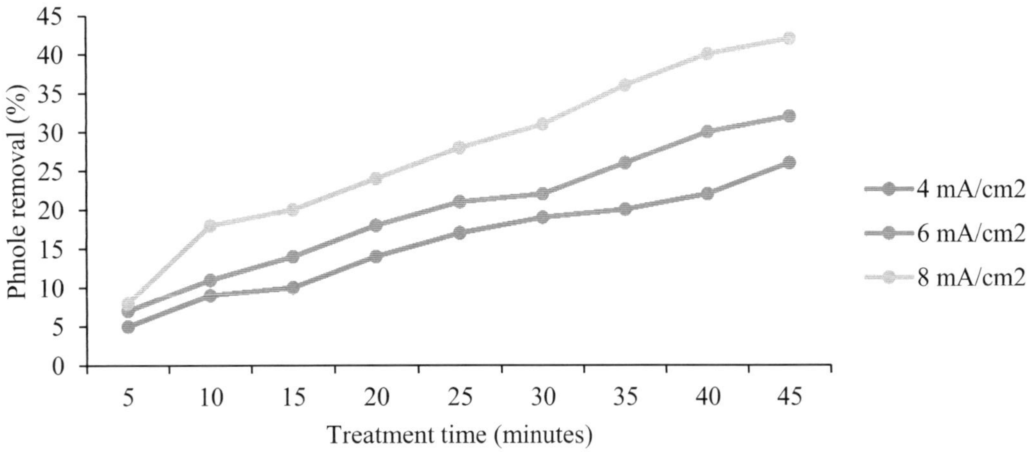

Figure 1: The effects of the ECD on the removal efficiency of the phenol using ER.

3.2. *Influence of AES*

The spacing between electrodes is another aspect that is heavily reported to affect the performance of the electrocoagulation method in terms of various pollutants treatment. Therefore, the effect of the effects between the aluminium electrodes has been examined in this research on the performance of the electrocoagulation in terms of phenols removal from refinery wastewater. The AES was changed manually using three different spacing that are 20, 25, and 30 mm. The remaining experimental variables which are ECD, initial concentration of the phenols and treatment time were kept on 4 mA/cm^2, 120 ppm, and 45 minutes, respectively. The results presented in figure 2 showed that the removal efficiency of the phenols using ER increases with the decrease in the AES. from figure 2 it can be seen that the removal efficiency of the phenols dropped from 27% to 15% when the AES increased from 20 mm to 30 mm. The change in the removal efficiency can be explained by the change in the electrostatic attraction force between the electrodes.

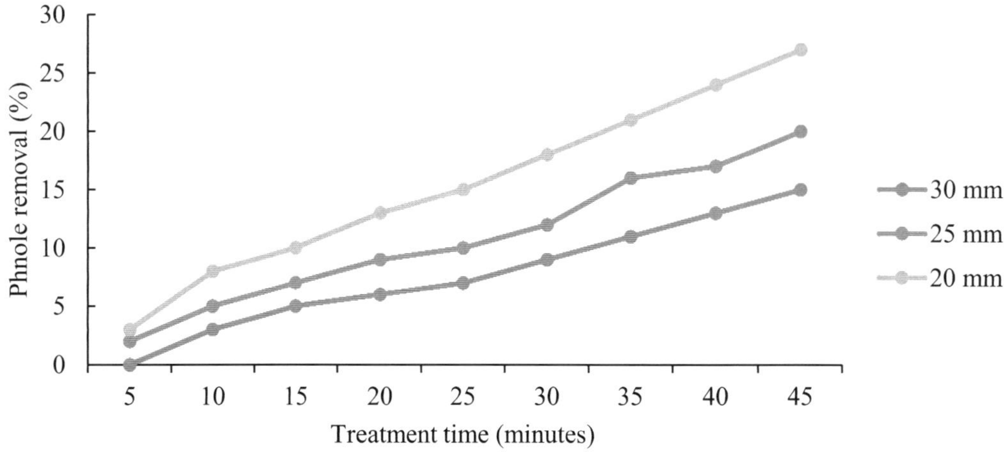

Figure 2: The effects of the AES on the removal efficiency of the phenol using ER

3.3. Influence of treatment time

It has been shown that the treatment time significantly affects the performance of various treatment technologies and method. in terms of electrocoagulation Abdulhadi, Kot, Hashim, Shaw, Muradov and Al-Khaddar [22] showed that the treatment time and the generation of coagulants are positively related. Based on this, the effects of treatment time on the removal of the phenols from refinery wastewater using the proposed ER has been studied in this research. To do this, samples that have phenolic pollutants concentration of 120 ppm was treated in the ER for 120 minutes using ECD of 6 mA/cm^2 and AES of 20 mm. From figure 3, the removal percentage of the phenols increased with time. phenols removal efficiency increased from 20% to 60% as the treating time increased from 20 minutes to 120 minutes.

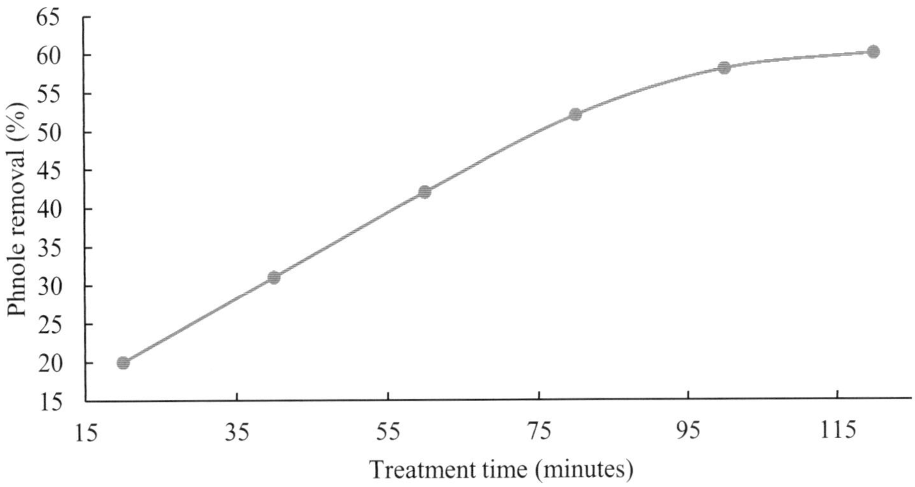

Figure 3: The effects of the treatment time on the removal efficiency of the phenol using ER.

4. Conclusion

This research investigated the application of the electrocoagulation method in terms of removing phenolic pollutants from petroleum refinery wastewater. Continuous flow experiments have been adopted to remove the pollutants. The outcome of this research showed that the presence of the phenol pollutants in refinery wastewater can significantly be reduced using the electrocoagulation method. Besides, it was found that the removal efficiency of the phenolic pollutants is influenced by the experimental parameters. The removal of the phenols increased with the decreases in the spacing between the electrodes while it was decreased with the decrease in the current density. Further researches are recommended to expel the effects of the initial concentration on the removal performance of the electrocoagulation method.

References

[1] Alattabi A W, Harris C, Alkhaddar R, Alzeyadi A and Abdulredha M J P e 2017 Online monitoring of a sequencing batch reactor treating domestic wastewater **196** 800-7

[2] Abdulredha M, Kadhim N, Hussein A, Almutairi M, Alkhaddar R, Yeboah D and Hashim K 2021 Zeolite as a natural adsorbent for nitrogenous compounds being removed from water. In: *IOP Conference Series: Materials Science and Engineering*: IOP Publishing) p 012082

[3] Jawad S F, Saddam N S, Adaami Q J, Kareem M M, Abdulredha M, Mubarak H A, Kot P, Gkantou M and AlKhayyat A 2021 Dye removal from textile wastewater using solar-powered electrocoagulation reactor. In: *IOP Conference Series: Materials Science and Engineering*: IOP Publishing) p 012016

[4] Aljuboury D, Palaniandy P, Abdul Aziz H and Feroz S 2017 Treatment of petroleum wastewater by conventional and new technologies-A review *Glob. Nest J* **19** 439-52

[5] Emamjomeh M M, Mousazadeh M, Mokhtari N, Jamali H A, Makkiabadi M, Naghdali Z, Hashim K S and Ghanbari R 2020 Simultaneous removal of phenol and linear alkylbenzene sulfonate from automotive service station wastewater: Optimization of coupled electrochemical and physical processes *Separation Science and Technology* **55** 3184-94

[6] Abdulhadi B A, Kot P, Hashim K S, Shaw A and Khaddar R A 2019 Influence of current density and electrodes spacing on reactive red 120 dye removal from dyed water using electrocoagulation/electroflotation (EC/EF) process. In: *First International Conference on Civil and Environmental Engineering Technologies (ICCEET),* (University of Kufa, Iraq pp 12-22

[7] Al-Saati N H, Hussein T K, Abbas M H, Hashim K, Al-Saati Z N, Kot P, Sadique M, Aljefery M H and Carnacina I 2019 Statistical modelling of turbidity removal applied to non-toxic natural coagulants in water treatment: a case study *Desalination and Water Treatment* **150** 406-12

[8] Hashim K S, Al-Saati N H, Hussein A H and Al-Saati Z N 2018 An investigation into the level of heavy metals leaching from canal-dreged sediment: a case study metals leaching from dreged sediment. In: *First International Conference on Materials Engineering & Science,* (Istanbul Aydın University (IAU), Turkey pp 12-22

[9] Diya'uddeen B H, Daud W M A W and Aziz A A 2011 Treatment technologies for petroleum refinery effluents: A review *Process safety and environmental protection* **89** 95-105

[10] Abdulredha M, Rafid A, Jordan D and Alattabi A J P e 2017 Facing up to waste: how can hotel managers in Kerbala, Iraq, help the city deal with its waste problem? **196** 771-8

[11] Abdulredha M, Muhsin A A, Al-Janabi A, Alajmi B N, Gkantou M, Amoako-Attah J, Al-Jumeily D, Mustafina J and AlKhayyat A 2021 Using SF and CKD as cement replacement materials for producing cement mortar. In: *IOP Conference Series: Materials Science and Engineering*: IOP Publishing) p 012007

[12] Isra'a S S, Al-Janabi A, Abdulredha M, Alkandari A, Abdellatif M and Yeboah D 2021 Reusing of furnace bottom ash as an adsorbent for phosphate removal from water. In: *IOP Conference Series: Materials Science and Engineering*: IOP Publishing) p 012006

[13] Hashim K S, Al-Saati N H, Alquzweeni S S, Zubaidi S L, Kot P, Kraidi L, Hussein A H, Alkhaddar R, Shaw A and Alwash R 2019 Decolourization of dye solutions by electrocoagulation: an investigation of the effect of operational parameters. In: *First International Conference on Civil and Environmental Engineering Technologies (ICCEET),* (University of Kufa, Iraq pp 25-32

[14] Hashim K S, Hussein A H, Zubaidi S L, Kot P, Kraidi L, Alkhaddar R, Shaw A and Alwash R 2019 Effect of initial pH value on the removal of reactive black dye from water by electrocoagulation (EC) method. In: *2nd International Scientific Conference,* (Al-Qadisiyah University, Iraq pp 12-22

[15] Hashim K S, Khaddar R A, Jasim N, Shaw A, Phipps D, Kot P, Pedrola M O, Alattabi A W, Abdulredha M and Alawsh R 2019 Electrocoagulation as a green technology for phosphate removal from River water *Separation and Purification Technology* **210** 135-44

[16] Hashim K S, AlKhaddar R, Shaw A, Kot P, Al-Jumeily D, Alwash R and Aljefery M H 2020 *Electrocoagulation as an eco-friendly River water treatment method. In Advances in Water Resources Engineering and Management* (Berline: Springer)

[17] Hassan Alnaimi I J I, Abuduljaleel Al-Janabi, Khalid Hashim, Michaela Gkantou, Salah L. Zubaidi, Patryk Kot, Magomed Muradov 2020 Ultrasonic-electrochemical treatment for effluents of concrete plants Ultrasonic-electrochemical treatment for effluents of concrete plants. In: *IOP Conference Series Materials Science and Engineering* (University of Kufa, Najaf, Iraq pp 1-9

[18] Abdulla G, Kareem M M, Hashim K S, Muradov M, Kot P, Mubarak H A, Abdellatif M and Abdulhadi B 2020 Removal of iron from wastewater using a hybrid filter. In: *IOP Conference Series: Materials Science and Engineering*: IOP Publishing) p 012035

[19] Omran I I, Al-Saati N H, Hashim K S, Al-Saati Z N, Patryk K, Khaddar R A, Al-Jumeily D, Shaw A, Ruddock F and Aljefery M 2019 Assessment of heavy metal pollution in the Great Al-Mussaib irrigation channel *Desalination and Water Treatment* **168** 165-74

[20] Abdulraheem F S, Al-Khafaji Z S, Hashim K S, Muradov M, Kot P and Shubbar A A 2020 Natural filtration unit for removal of heavy metals from water. In: *IOP Conference Series: Materials Science and Engineering*: IOP Publishing) p 012034

[21] Alyafei A, AlKizwini R S, Hashim K S, Yeboah D, Gkantou M, Al Khaddar R, Al-Faluji D and Zubaidi S L 2020 Treatment of effluents of construction industry using a combined filtration-electrocoagulation method. In: *IOP Conference Series: Materials Science and Engineering*: IOP Publishing) p 012032

[22] Abdulhadi B, Kot P, Hashim K, Shaw A, Muradov M and Al-Khaddar R 2021 Continuous-flow electrocoagulation (EC) process for iron removal from water: Experimental, statistical and economic study *Science of The Total Environment* **756** 1-16

[23] Mohammed A-H, Hussein A H, Yeboah D, Al Khaddar R, Abdulhadi B, Shubbar A A and Hashim K S 2020 Electrochemical removal of nitrate from wastewater. In: *IOP Conference Series: Materials Science and Engineering*: IOP Publishing) p 012037

[24] Hashim K, Kot P, Zubaid S, Alwash R, Al Khaddar R, Shaw A, Al-Jumeily D and Aljefery M 2020 Energy efficient electrocoagulation using baffle-plates electrodes for efficient Escherichia Coli removal from Wastewater *Journal of Water Process Engineering* **33** 101079-86

[25] Alhendal M, Nasir M J, Hashim K S, Amoako-Attah J, Al-Faluji D, Muradov M, Kot P and Abdulhadi B 2020 Cost-effective hybrid filter for remediation of water from fluoride. In: *IOP Conference Series: Materials Science and Engineering*: IOP Publishing) p 012038

[26] Al-Marri S, AlQuzweeni S S, Hashim K S, AlKhaddar R, Kot P, AlKizwini R S, Zubaidi S L and Al-Khafaji Z S 2020 Ultrasonic-Electrocoagulation method for nitrate removal from water. In: *IOP Conference Series: Materials Science and Engineering*: IOP Publishing) p 012073

[27] Alenezi A K, Hasan H A, Hashim K S, Amoako-Attah J, Gkantou M, Muradov M, Kot P and Abdulhadi B 2020 Zeolite-assisted electrocoagulation for remediation of phosphate from calcium-phosphate solution. In: *IOP Conference Series: Materials Science and Engineering*: IOP Publishing) p 012031

[28] Hashim K S, Ewadh H M, Muhsin A A, Zubaidi S L, Kot P, Muradov M, Aljefery M and Al-Khaddar R 2020 Phosphate removal from water using bottom ash: Adsorption performance, coexisting anions and modelling studies *Water Science and Technology* **83** 1-17

[29] Hashim K S, Idowu I A, Jasim N, Al Khaddar R, Shaw A, Phipps D, Kot P, Pedrola M O, Alattabi A W and Abdulredha M 2018 Removal of phosphate from River water using a new baffle plates electrochemical reactor *MethodsX* **5** 1413-8

[30] Hashim K S, Ali S S M, AlRifaie J K, Kot P, Shaw A, Al Khaddar R, Idowu I and Gkantou M 2020 Escherichia coli inactivation using a hybrid ultrasonic–electrocoagulation reactor *Chemosphere* **247** 125868-75

[31] Hashim K S, Shaw A, AlKhaddar R, Kot P and Al-Shamma'a A 2021 Water purification from metal ions in the presence of organic matter using electromagnetic radiation-assisted treatment *Journal of Cleaner Production* **280**

[32] Zanki A K, Mohammad F H, Hashim K S, Muradov M, Kot P, Kareem M M and Abdulhadi B 2020 Removal of organic matter from water using ultrasonic-assisted electrocoagulation method. In: *IOP Conference Series: Materials Science and Engineering*: IOP Publishing) p 012033

[33] Alattabi A W, Harris C, Alkhaddar R, Alzeyadi A and Hashim K 2017 Treatment of Residential Complexes' Wastewater using Environmentally Friendly Technology *Procedia Engineering* **196** 792-9

[34] Alattabi A W, Harris C B, Alkhaddar R M, Hashim K S, Ortoneda-Pedrola M and Phipps D 2017 Improving sludge settleability by introducing an innovative, two-stage settling sequencing batch reactor *Journal of Water Process Engineering* **20** 207-16

[35] Aqeel K, Mubarak H A, Amoako-Attah J, Abdul-Rahaim L A, Al Khaddar R, Abdellatif M, Al-Janabi A and Hashim K S 2020 Electrochemical removal of brilliant green dye from wastewater. In: *IOP Conference Series: Materials Science and Engineering*: IOP Publishing) p 012036

[36] Hashim K S, Shaw A, Al Khaddar R, Ortoneda Pedrola M and Phipps D 2017 Defluoridation of drinking water using a new flow column-electrocoagulation reactor (FCER) - Experimental, statistical, and economic approach *Journal of Environmental Management* **197** 80-8

[37] Hashim K S, Shaw A, Al Khaddar R, Pedrola M O and Phipps D 2017 Iron removal, energy consumption and operating cost of electrocoagulation of drinking water using a new flow column reactor *Journal of Environmental Management* **189** 98-108

[38] Hashim K S, Shaw A, Al Khaddar R, Pedrola M O and Phipps D 2017 Energy efficient electrocoagulation using a new flow column reactor to remove nitrate from drinking water - Experimental, statistical, and economic approach *Journal of Environmental Management* **196** 224-33

[39] Shubbar A A, Jafer H, Dulaimi A, Hashim K, Atherton W and Sadique M 2018 The development of a low carbon binder produced from the ternary blending of cement, ground granulated blast furnace slag and high calcium fly ash: An experimental and statistical approach *Construction and Building Materials* **187** 1051-60

[40] Shubbar A A, Sadique M, Nasr M S, Al-Khafaji Z S and Hashim K S 2020 The impact of grinding time on properties of cement mortar incorporated high volume waste paper sludge ash *Karbala International Journal of Modern Science* **6** 1-23

[41] Shubbar A A, Sadique M, Shanbara H K and Hashim K 2020 *The Development of a New Low Carbon Binder for Construction as an Alternative to Cement. In Advances in Sustainable Construction Materials and Geotechnical Engineering* (Berlin: Springer)

[42] Idowu I A, Atherton W, Hashim K, Kot P, Alkhaddar R, Alo B I and Shaw A 2019 An analyses of the status of landfill classification systems in developing countries: Sub Saharan Africa landfill experiences *Waste Management* **87** 761-71

[43] Al-Jumeily D, Hashim K, Alkaddar R, Al-Tufaily M and Lunn J 2019 Sustainable and Environmental Friendly Ancient Reed Houses (Inspired by the Past to Motivate the Future). In: *11th International Conference on Developments in eSystems Engineering (DeSE),* (Cambridge, UK pp 214-9

[44] Abdulredha M, Rafid A, Jordan D and Hashim K 2017 The development of a waste management system in Kerbala during major pilgrimage events: determination of solid waste composition *Procedia Engineering* **196** 779-84

[45] Abdulredha M, Al Khaddar R, Jordan D, Kot P, Abdulridha A and Hashim K 2018 Estimating solid waste generation by hospitality industry during major festivals: A quantification model based on multiple regression *Waste Management* **77** 388-400

[46] Abdulredha M, Rafid A, Jordan D and Hashim K J P E 2017 The development of a waste management system in Kerbala during major pilgrimage events: determination of solid waste composition **196** 779-84

[47] Abdulredha M, Al Khaddar R, Jordan D, Kot P, Abdulridha A and Hashim K J W M 2018 Estimating solid waste generation by hospitality industry during major festivals: A quantification model based on multiple regression **77** 388-400

[48] Abdulredha M, al-Khaddar R, Kot P, Jordan D and Abdulridha A 2018 Benchmarking of the Current Solid Waste Management System in Karbala, Iraq, Using Wasteaware Benchmark Indicators. In: *World Environmental and Water Resources Congress 2018: Groundwater, Sustainability, and Hydro-Climate/Climate Change*: American Society of Civil Engineers Reston, VA) pp 40-8

[49] Abdulredha M, Abdulridha A, Shubbar A, Alkhaddar R, Kot P and Jordan D 2020 Estimating municipal solid waste generation from service processions during the Ashura religious event. In: *IOP Conference Series: Materials Science and Engineering*: IOP Publishing) p 012075
[50] Hashim K S, Idowu I A, Jasim N, Al Khaddar R, Shaw A, Phipps D, Kot P, Pedrola M O, Alattabi A W and Abdulredha M J M 2018 Removal of phosphate from River water using a new baffle plates electrochemical reactor **5** 1413-8
[51] Majdi H S, Shubbar A, Nasr M S, Al-Khafaji Z S, Jafer H, Abdulredha M, Al Masoodi Z, Sadique M and Hashim K J D i B 2020 Experimental data on compressive strength and ultrasonic pulse velocity properties of sustainable mortar made with high content of GGBFS and CKD combinations

Phosphate removal from wastewater by furnace bottom ash

D Mahmood [1]
[1] Faculty of Science, Mustansiriyah University, Iraq.

Email: drmahmood90@gmaol.com

Abstract. Orthophosphate is a natural chemical found in large quantities in the Earth's crust, and its inclusion in water causes eutrophication. As a result, several experiments have been conducted to investigate the potential of various treatment systems to remove phosphate from wastewater. According to recent research, filtration methods are useful for phosphate removal, with one drawback being the expensive cost of the filtration materials. Therefore, new research has focused on reducing the cost of filtration method using cheap replacements. The current research focuses on the efficacy of using iron industry wastes which the blast furnace bottom ashes to remove phosphates from contaminated wastewater. Some operating conditions, like and phosphate level, treating periods and blast furnace bottom ashes dosages, have investigated in order to find a suitable set for achieving the highest probable removal rate at a fair charge. The findings showed that the bottom ashes are an effective phosphate removal substitute. The highest removal of phosphate was 91.2% at 37.5 minutes treatment period, using initial concertation of 5 ppm, and a dose of the bottom ashes of 575 ppm. The findings have been used to build a production model with a high level of reliability.

Keyword: Blast furnace bottom ashes, phosphate; wastewater.

1. Introduction

The demand for fresh potable water is growing all over the world as a result of the rapid growth of population growth and dwindling clean water supplies [1-3]. Since the discharged wastewater volume is measured in strict proportion to the population, the increasing population increases the quantity of wastewater disposed to waterbodies and turn the negative consequences on the environment and human [1, 3-5]. As a result, overt discharge of toxins into freshwater should be banned, and drainage should be properly handled, whether residential, industrial, and agricultural effluents, in order to remove or reduce the harmful effects of the pollutants [6, 7]. Algae flourishing is caused by phosphorus infiltration of water, which harms water.

Phosphate is considered an important source of nutrients and, therefore, is considered a common pollutant in all kinds of wastewater effluents [8-11]. The biggest problem produced by the pollution of phosphate is the rapid growth of algae which absorbing sunshine and depleting the oxygen in water bodies. This could result in the death of a significant number of water animals and plants in a brief period [12-14]. Because of the dramatic increase in organic matter concentration, the water's flavour, odour, and composition will be greatly affected. Algal growth at an increased rate could have devastating consequences for human health, manufacturing and fisheries and farmers. Many researchers discovered that phytoplankton species can produce toxic materials that cause a variety of harmful diseases and, in some cases, death [15-21]. The Environmental Protection Agency has specific guidelines when it comes to the maximum permissible pollutant concentrations in water. The permissible phosphate concertation in drinking water is set to be 0.1 ppm, while it is 0.05 ppm in lakes and rivers, according to the Environmental Protection Agency. Based on the above, it is essential to remove the phosphate from polluted water. However, using conventional procedures to remove phosphate usually necessitates an expensive and time-consuming treatment process [12, 22]. To give an example, the standard filtration process, with certain improvements, provide a promising method for phosphate. However, this method is considered a non-economical method due to the use of costly filters. Therefore, the focus of this research was on the potential to use furnaces bottom ashes as industrial waste as a low-cost adsorbent to remove phosphate from polluted wastewater.

2. Review of literature

Phosphorus elimination from the aquatic environment is accomplished by changing the format of the element to a solid format, which is then removed from the solution through floatation or other methods [12, 23]. Phosphate is converted using a variety of methods, including chemical treatment, ponds, biological treatment, adsorption, etc.

The biological treatment of phosphate can be performed by using the normal actions of various microorganisms such as microalgae and bacteria. These methods can be anaerobic or aerobic, based on the type of microorganism. The treatment effectiveness is depending on various aspects like pH, initial concentration etc.[24-30]. Many drawbacks are found in terms of using biomedical treatment to eliminate the presence of phosphate. This includes the presence of ample biodegradable carbon, the rapid growth of microorganism, and the pH level of the solution as pH changes could destroy the bacteria [31-38]. Chemical techniques involve adding metallic salts to phosphate contaminated water to convert phosphate ions. Chemical methods are commonly used for the treatment of wastewater containing phosphate [39-43]. Because of its obtainability, cheapness, and ease of use, calcium ion is regarded as one of the most used chemical additives for phosphate treatment. The aluminium hydroxide is considered very efficient in removing phosphate from contaminated water, particularly in an acidic environment. Chemical treatment processes often result in poisonous sludge because of the use of chemical materials. An additional consideration is that the procedure is more expensive than most processes due to the use of chemicals [44, 45]. Physical phosphate removal techniques do not alter the chemical structure of the phosphate which are usually reversible. Based on this method the removed phosphorus pollutants may be isolated from the removing media. Membrane and filtration and membrane methods are widely used physical techniques for phosphate elimination [46, 47].

Filtration is an extraction technique that employs permeable media to remove many contaminants. Following that, various techniques like precipitation, flocculation, impaction, adsorption, and straining are used to separate contaminants from water. To eliminate phosphate from water, the filtration process used various permeable media [5, 48-51]. Overall, filtration cost is dictated by the cost of used adsorption materials, which are typically expensive particularly for industrial materials. Earlier experiments on membrane approaches revealed that membrane methods are a strong capacity to remove phosphate. Reverse osmosis is one of the wildly adopted methods to remove low phosphate concentrations (below 0.008 mg/L) [3, 52, 53]. Reverse osmosis is based on the pressure-driven way of contaminated solution in a membrane with very low permeability. The member helps to remove the contaminants from water by pressure. But, because of the high operating energy and costs, the difficulty of disposing of the membrane concentrators, and the need for specialized apparatus and skills, membrane system is uncommonly used. As a result, recent research has centred on the adoption of industrial waste or environmental resources as efficient and functional substitutes. Fly ash is a common expel of industrial waste in phosphate removal from water. Research showed that this material is effective in removing the phosphate from wastewater that has pH level of 7 [20, 21, 30].

3. Research methods

As stated earlier, the current research aimed to see whether blast furnace bottom ashes could be used as cheap and adsorbent to remove phosphate from. It's crucial to study the physical properties of the bottom ashes like size distribution, porosity, surface areas and specific gravity. It is very important to conduct the particle size distribution since it is linked to the area available for adsorption. Sieve numbers of 4,6,10,16,30, and 50 are used conducted using sieve analysis, and the quantity of materials retained on every mesh is determined to examine the particle size distribution of the waste. Besides, the following equation is used to find the specific gravity (SG) test using the gas jar technique:

$$SG = \frac{Y_2 - Y_1}{(Y_4 - Y_1) - (Y_3 - Y_2)} \quad (1)$$

Where

Y_1: Jar weight + plate weight

Y_2: The combined weight of the sample, plate, and jar.

Y_3: The combined weight of the water, dray sample, plate, and jar.

Y_4: Jar weight + water

Following this, a chemical test was performed to see the composition of the furnaces bottom ashes using an X-ray fluorescence analyzer. The chemical composition test is done to determine the removing efficiency of the furnaces bottom ashes because it has been discovered that compounds containing calcium, iron, and aluminium, can be used for phosphate treatment. The bottom ashes were used to eliminate the presence of phosphate in polluted water samples. In terms of furnaces bottom ashes dose, treatment period and temperature, the central composite configuration was used to optimize the method. The test implemented various concentrations of phosphate, which were modified and regulated by combining potassium diphosphate within deionized water.

To investigate the impact of the time, the contaminated water was mixed with the furnaces bottom ashes were kept mixed for varying amounts of time in one bottle. In these experiments, the detention period was 10 to 65 minutes, the furnaces bottom ashes doses ranged from 150 to 750 mg/L and the concentration of the phosphate ranged from 5 to 15 mg/L. Hach Lange spectrophotometer is adopted to calculate the concentrations of the phosphate.

The central composite design has been used in this research to build the production model in order to attain the best water treatment as this technique is an effective and largely adopted method. Phosphate elimination was the response variable in this study. Phosphate concentration level, furnaces bottom ashes dosage, and treatment duration were the predictors. Minitab 19 has been adopted to create the production matrix. Table 1 shows the independent variable ranges that were used.

Table 1. The independent and dependent variable ranges that were investigated.

Independent variables	units	Code	Level		
			Low	Moderate	Elevated
Phosphate concentration level	mg/L	PCL	5.0	10.0	15.0
Furnaces bottom ashes dosage	mg/L	FBADs	150.0	575.0	1000.0
Treatment duration	Minutes	TD	10.0	37.5	65.0

Following this, the α value, which creates a useful predictor of linearity and indicated significance is determined using the following equation:

$$\alpha = \sqrt[4]{2a} \quad (2)$$

Where a represents the number of used predictors.

4. Results

4.1. *Furnaces bottom ashes properties*

Most of the bottom ashes particles are retained on sieves of smaller sizes (sieves 4 to 16) based on the experimental results. The outcome of the sieves analyses suggests that small particles sizes between 1.180mm and 4.750 mm made up the bulk of the bottom ashes sample. The level of gradation was calculated using the below equations (3, and 4):

$$\text{Uniformity coefficient} = \frac{D_{60}}{D_{10}} \quad (3)$$

$$\text{The gradation coefficient} = \frac{D_{30}^2}{D_{10} \times D_{60}} \quad (4)$$

Where D_{60}, D_{30}, and D_{10} represents the percentage of particles passing on each noted size while were found for the plot of the passing rate with particle size. The results of the analyses showed that the D_{60}, D_{30}, and D_{10} values were 3.08 1.40, and 0.50, respectively. In addition, the gradation coefficient is 1.27 and the uniformity coefficient is 6.16. the bottom ashes samples are deemed well-graded. Besides, the specific gravity of the bottom ashes was found to be 1.25 which means that the ashes will not float as it has specific gravity is larger than water specific gravity. Additionally, the permeability of the ashes was found to be 0.58 m^2/g, and the surface area was 6.2 m^2/g. based on these results the furnaces bottom ashes can be considered ideal for phosphate removal from water, as they have a large area, proper permeability, and higher specific gravity.

On the other hand, based on the chemical tests, it was found that about 20% of the furnaces bottom ashes is calcium oxides, magnesium, iron, and aluminium. Thereby, it can be said that the furnace bottom ashes, is ideal for phosphate removal.

4.2. *Phosphate treatment*

In this research batch tests were used to assess the removal of the phosphate from water using furnace bottom ashes. 20 trials where experimental variables were changed in each trial to determine the best experimental set to achieve high removal efficiency of low costs. Table 2 displays the results. The central composite design and removal efficiency achieved is presented in Table 2.

Table 2. Central composite design for phosphate removal in a batch flow experiment.

Trail	PCL	FBADs	TD	Removal %
01	-1	-1	-1	72.1
02	1	-1	-1	83.5
03	-1	1	-1	77.2
04	1	1	-1	65.7
05	-1	-1	1	63.1
06	1	-1	1	58.2
07	-1	1	1	75.2
08	1	1	1	79.1
09	-1.682	0	0	59.9
10	1.682	0	0	80.1
11	0	-1.682	0	56.2
12	0	1.682	0	77.3
13	0	0	-1.682	91.2
14	0	0	1.682	66.4
15	0	0	0	70.1
16	0	0	0	79.2
17	0	0	0	78.1
18	0	0	0	79.2
19	0	0	0	77.8
20	0	0	0	78.1

The findings of this research study revealed that the treatment efficiency is related to the dosage of the furnace's bottom ashes and treatment duration. However, phosphate treatment has a negative relationship with the initial concentration the phosphate. The available surface area of the adsorbent is not sufficient to absorb the phosphate from the water, resulting in a decreased treatment performance. Implementing a higher concentration of furnace's bottom ashes increases the available area for adsorption of the phosphate from polluted water. The increased treatment tie increases the contact between the adsorbent and the phosphate. Using 75 mg/L of furnace's bottom ashes for 37.5 minutes were appropriate to remove 91.2 of the phosphate concertation in the polluted water as presented in Figure 1.

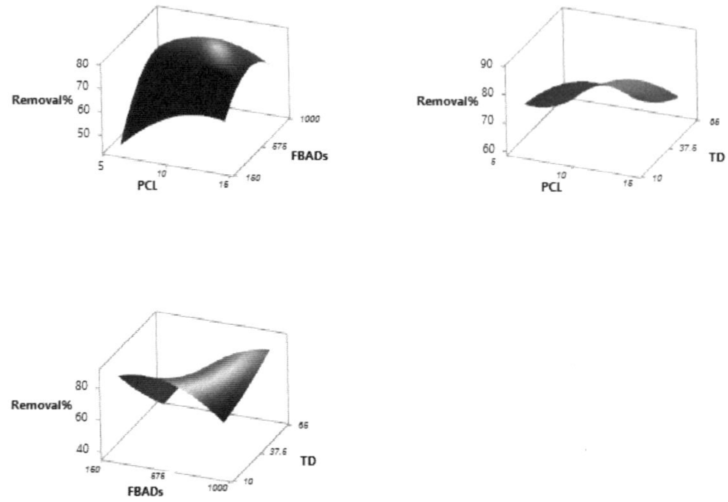

Figure 1: surface plots of the removal performance of phosphate vs experimental parameters

Equation 5 were developed using the central composite design method to estimate the removal efficiency based on the used parameters.

$$Removal\,\% = 77.08 + 2.41\,\text{PCL} + 4.08\,\text{FBADs} - 4.73\,\text{TD} - 2.46\,\text{PCL} * \text{PCL} - 3.60\,\text{FBADs} * \text{FBADs} + 0.66\,\text{TD} * \text{TD} - 1.76\,\text{PCL} * \text{FBADs} - 0.11\,\text{PCL} * \text{TD} + 5.71\,\text{FBADs} * \text{TD} \quad (5)$$

5. Conclusion

The potential to use furnaces bottom ashes to treat the presences of phosphate from polluted was investigated in order to determine the best operating considerations. Research findings showed that furnaces bottom ashes chemical and physical characteristics are necessary to remove phosphate from polluted water. The performance of the furnaces bottom ashes is like previous research finding of around 90% per cent removal. The treatment percentage is proportional to the dosage of the furnace's bottom ashes and treatment duration. On the other hand, the removal efficiency is negatively connected to phosphate removal. In addition, the developed central composite model can predict the change in the removal efficiency of the phosphate-based on the dose of the furnace's bottom ashes, the concentration of the phosphate and the treatment time.

References

[1] Alenezi A K, Hasan H A, Hashim K S, Amoako-Attah J, Gkantou M, Muradov M, Kot P and Abdulhadi B 2020 Zeolite-assisted electrocoagulation for remediation of phosphate from calcium-phosphate solution. In: *IOP Conference Series: Materials Science and Engineering*: IOP Publishing) p 012031

[2] Hashim K S, Khaddar R A, Jasim N, Shaw A, Phipps D, Kot P, Pedrola M O, Alattabi A W, Abdulredha M and Alawsh R 2019 Electrocoagulation as a green technology for phosphate removal from River water *Separation and Purification Technology* **210** 135-44

[3] Jawad S F, Saddam N S, Adaami Q J, Kareem M M, Abdulredha M, Mubarak H A, Kot P, Gkantou M and AlKhayyat A 2021 Dye removal from textile wastewater using

solar-powered electrocoagulation reactor. In: *IOP Conference Series: Materials Science and Engineering*: IOP Publishing) p 012016

[4] Hashim K S, Al-Saati N H, Alquzweeni S S, Zubaidi S L, Kot P, Kraidi L, Hussein A H, Alkhaddar R, Shaw A and Alwash R 2019 Decolourization of dye solutions by electrocoagulation: an investigation of the effect of operational parameters. In: *First International Conference on Civil and Environmental Engineering Technologies (ICCEET),* (University of Kufa, Iraq pp 25-32

[5] Isra'a S S, Al-Janabi A, Abdulredha M, Alkandari A, Abdellatif M and Yeboah D 2021 Reusing of furnace bottom ash as an adsorbent for phosphate removal from water. In: *IOP Conference Series: Materials Science and Engineering*: IOP Publishing) p 012006

[6] Hashim K S, Idowu I A, Jasim N, Al Khaddar R, Shaw A, Phipps D, Kot P, Pedrola M O, Alattabi A W and Abdulredha M 2018 Removal of phosphate from River water using a new baffle plates electrochemical reactor *MethodsX* **5** 1413-8

[7] Al-Jumeily D, Hashim K, Alkaddar R, Al-Tufaily M and Lunn J 2019 Sustainable and Environmental Friendly Ancient Reed Houses (Inspired by the Past to Motivate the Future). In: *11th International Conference on Developments in eSystems Engineering (DeSE),* (Cambridge, UK pp 214-9

[8] Hashim K S, Shaw A, Al Khaddar R, Pedrola M O and Phipps D 2017 Energy efficient electrocoagulation using a new flow column reactor to remove nitrate from drinking water - Experimental, statistical, and economic approach *Journal of Environmental Management* **196** 224-33

[9] Abdulhadi B, Kot P, Hashim K, Shaw A, Muradov M and Al-Khaddar R 2021 Continuous-flow electrocoagulation (EC) process for iron removal from water: Experimental, statistical and economic study *Science of The Total Environment* **756** 1-16

[10] Hashim K S, Hussein A H, Zubaidi S L, Kot P, Kraidi L, Alkhaddar R, Shaw A and Alwash R 2019 Effect of initial pH value on the removal of reactive black dye from water by electrocoagulation (EC) method. In: *2nd International Scientific Conference,* (Al-Qadisiyah University, Iraq pp 12-22

[11] Hashim K S, Ewadh H M, Muhsin A A, Zubaidi S L, Kot P, Muradov M, Aljefery M and Al-Khaddar R 2020 Phosphate removal from water using bottom ash: Adsorption performance, coexisting anions and modelling studies *Water Science and Technology* **83** 1-17

[12] Hashim K S, AlKhaddar R, Shaw A, Kot P, Al-Jumeily D, Alwash R and Aljefery M H 2020 *Electrocoagulation as an eco-friendly River water treatment method. In Advances in Water Resources Engineering and Management* (Berline: Springer)

[13] Zubaidi S L, Ortega-Martorell S, Kot P, Alkhaddar R M, Abdellatif M, Gharghan S K, Ahmed M S and Hashim K 2020 A Method for Predicting Long-Term Municipal Water Demands Under Climate Change *Water Resources Management* **34** 1265-79

[14] Hashim K S, Shaw A, AlKhaddar R, Kot P and Al-Shamma'a A 2021 Water purification from metal ions in the presence of organic matter using electromagnetic radiation-assisted treatment *Journal of Cleaner Production* **280**

[15] Abdulredha M, Kot P, Al Khaddar R, Jordan D, Abdulridha A J E, Development and Sustainability 2020 Investigating municipal solid waste management system performance during the Arba'een event in the city of Kerbala, Iraq **22** 1431-54

[16] Abdulredha M, Muhsin A A, Al-Janabi A, Alajmi B N, Gkantou M, Amoako-Attah J, Al-Jumeily D, Mustafina J and AlKhayyat A 2021 Using SF and CKD as cement replacement materials for producing cement mortar. In: *IOP Conference Series: Materials Science and Engineering*: IOP Publishing) p 012007

[17] Abdulredha M, Rafid A, Jordan D and Alattabi A J P e 2017 Facing up to waste: how can hotel managers in Kerbala, Iraq, help the city deal with its waste problem? **196** 771-8

[18] Abdulredha M, Rafid A, Jordan D and Hashim K J P E 2017 The development of a waste management system in Kerbala during major pilgrimage events: determination of solid waste composition **196** 779-84

[19] Al-Anbari R, Alnakeeb A, Abdulredha M J E and Journal T 2013 Landfill site selection for Kerbala municipal solid wastes by using geographical information system techniques **32** 13

[20] Alattabi A W, Harris C, Alkhaddar R, Alzeyadi A and Abdulredha M J P e 2017 Online monitoring of a sequencing batch reactor treating domestic wastewater **196** 800-7

[21] Al-Sareji O J, Abdulredha M, Mubarak H A, Grmasha R A, Alnowaishry A, Kot P, Al-Khaddar R and AlKhayyat A 2021 Copper removal from water using carbonized sawdust. In: *IOP Conference Series: Materials Science and Engineering*: IOP Publishing) p 012015

[22] Zanki A K, Mohammad F H, Hashim K S, Muradov M, Kot P, Kareem M M and Abdulhadi B 2020 Removal of organic matter from water using ultrasonic-assisted electrocoagulation method. In: *IOP Conference Series: Materials Science and Engineering*: IOP Publishing) p 012033

[23] Al-Saati N H, Hussein T K, Abbas M H, Hashim K, Al-Saati Z N, Kot P, Sadique M, Aljefery M H and Carnacina I 2019 Statistical modelling of turbidity removal applied to non-toxic natural coagulants in water treatment: a case study *Desalination and Water Treatment* **150** 406-12

[24] Aayef A N, Al Masoodi W T M, Kamel R J, Abdulredha M, Almansoori N A, Kot P and Muradov M 2021 An experimental study for adapting electrocoagulation as a technique for fluoride removal from water. In: *IOP Conference Series: Materials Science and Engineering*: IOP Publishing) p 012012

[25] Abdulredha M, Abdulridha A, Shubbar A, Alkhaddar R, Kot P and Jordan D 2020 Estimating municipal solid waste generation from service processions during the Ashura religious event. In: *IOP Conference Series: Materials Science and Engineering*: IOP Publishing) p 012075

[26] Abdulredha M, Al Khaddar R and Jordan D 2017 Hoteliers' attitude towards solid waste source separation through mega festivals: A pilot study in Karbala. In: *International Conference for Doctoral Research*: BUID)

[27] Abdulredha M, Al Khaddar R, Jordan D, Al-Attabi A and Alzeyadi A 2017 Public participation in solid waste management during mega festivals: A pilot study. In: *WCST World Congress on Sustainable Technologies Proceedings 2017*: Infonomics Society) pp 38-41

[28] Abdulredha M, Al Khaddar R, Jordan D, Kot P, Abdulridha A and Hashim K J W M 2018 Estimating solid waste generation by hospitality industry during major festivals: A quantification model based on multiple regression **77** 388-400

[29] Abdulredha M, al-Khaddar R, Kot P, Jordan D and Abdulridha A 2018 Benchmarking of the Current Solid Waste Management System in Karbala, Iraq, Using Wasteaware Benchmark Indicators. In: *World Environmental and Water Resources Congress 2018: Groundwater, Sustainability, and Hydro-Climate/Climate Change*: American Society of Civil Engineers Reston, VA) pp 40-8

[30] Abdulredha M, Kadhim N, Hussein A, Almutairi M, Alkhaddar R, Yeboah D and Hashim K 2021 Zeolite as a natural adsorbent for nitrogenous compounds being removed from water. In: *IOP Conference Series: Materials Science and Engineering*: IOP Publishing) p 012082

[31] Hassan Alnaimi I J I, Abuduljaleel Al-Janabi, Khalid Hashim, Michaela Gkantou, Salah L. Zubaidi, Patryk Kot, Magomed Muradov 2020 Ultrasonic-electrochemical treatment for effluents of concrete plants Ultrasonic-electrochemical treatment for effluents of concrete plants. In: *IOP Conference Series Materials Science and Engineering* (University of Kufa, Najaf, Iraq pp 1-9

[32] Majdi H S, Shubbar A, Nasr M S, Al-Khafaji Z S, Jafer H, Abdulredha M, Masoodi Z A, Sadique M and Hashim K 2020 Experimental data on compressive strength and ultrasonic pulse velocity properties of sustainable mortar made with high content of GGBFS and CKD combinations *Data in Brief* **31** 105961-72

[33] Kadhim A, Sadique M, Al-Mufti R and Hashim K 2020 Developing One-Part Alkali-Activated metakaolin/natural pozzolan Binders using Lime Waste as activation Agent *Advances in Cement Research* **32** 1-38

[34] Shubbar A A, Jafer H, Dulaimi A, Hashim K, Atherton W and Sadique M 2018 The development of a low carbon binder produced from the ternary blending of cement, ground granulated blast furnace slag and high calcium fly ash: An experimental and statistical approach *Construction and Building Materials* **187** 1051-60

[35] Shubbar A A, Al-Shaer A, AlKizwini R S, Hashim K, Hawesah H A and Sadique M 2019 Investigating the influence of cement replacement by high volume of GGBS and PFA on the mechanical performance of cement mortar. In: *First International Conference on Civil and Environmental Engineering Technologies (ICCEET),* (University of Kufa, Iraq pp 31-8

[36] Zubaidi S L, Gharghan S K, Dooley J, Alkhaddar R M and Abdellatif M 2018 Short-Term Urban Water Demand Prediction Considering Weather Factors *Water Resources Management* **32** 4527-42

[37] Farhan S, Akef V and Nasar Z 2020 The transformation of the inherited historical urban and architectural characteristics of Al-Najaf's Old City and possible preservation insights *Frontiers of Architectural Research*

[38] Farhan S L and Nasar A Z 2021 Urban Identity in the Pilgrimage Cities of Iraq: Analysis Trends of Architectural Designers in the City of Karbala *J. Urban Regen. Renew* **14** 2-14

[39] Shubbar A A, Sadique M, Shanbara H K and Hashim K 2020 *The Development of a New Low Carbon Binder for Construction as an Alternative to Cement. In Advances in Sustainable Construction Materials and Geotechnical Engineering* (Berlin: Springer)

[40] Shubbar A A, Sadique M, Nasr M S, Al-Khafaji Z S and Hashim K S 2020 The impact of grinding time on properties of cement mortar incorporated high volume waste paper sludge ash *Karbala International Journal of Modern Science* **6** 1-23

[41] Abdulhadi B A, Kot P, Hashim K S, Shaw A and Khaddar R A 2019 Influence of current density and electrodes spacing on reactive red 120 dye removal from dyed water using electrocoagulation/electroflotation (EC/EF) process. In: *First International Conference on Civil and Environmental Engineering Technologies (ICCEET),* (University of Kufa, Iraq pp 12-22

[42] Abdulredha M, Rafid A, Jordan D and Alattabi A 2017 Facing up to waste: how can hotel managers in Kerbala, Iraq, help the city deal with its waste problem? *Procedia engineering* **196** 771-8

[43] Alattabi A W, Harris C, Alkhaddar R, Alzeyadi A and Abdulredha M 2017 Online Monitoring of a sequencing batch reactor treating domestic wastewater *Procedia engineering* **196** 800-7

[44] Al-Marri S, AlQuzweeni S S, Hashim K S, AlKhaddar R, Kot P, AlKizwini R S, Zubaidi S L and Al-Khafaji Z S 2020 Ultrasonic-Electrocoagulation method for nitrate removal from water. In: *IOP Conference Series: Materials Science and Engineering*: IOP Publishing) p 012073

[45] Alhendal M, Nasir M J, Hashim K S, Amoako-Attah J, Al-Faluji D, Muradov M, Kot P and Abdulhadi B 2020 Cost-effective hybrid filter for remediation of water from fluoride. In: *IOP Conference Series: Materials Science and Engineering*: IOP Publishing) p 012038

[46] Aqeel K, Mubarak H A, Amoako-Attah J, Abdul-Rahaim L A, Al Khaddar R, Abdellatif M, Al-Janabi A and Hashim K S 2020 Electrochemical removal of brilliant green dye from wastewater. In: *IOP Conference Series: Materials Science and Engineering*: IOP Publishing) p 012036

[47] Alyafei A, AlKizwini R S, Hashim K S, Yeboah D, Gkantou M, Al Khaddar R, Al-Faluji D and Zubaidi S L 2020 Treatment of effluents of construction industry using a combined filtration-electrocoagulation method. In: *IOP Conference Series: Materials Science and Engineering*: IOP Publishing) p 012032

[48] ALWAN H H, SALEH L A, AL-MOHAMMED F M, ABDULREDHA M A J J o E S and Technology 2020 EXPERIMENTAL PREDICTION OF THE DISCHARGE COEFFICIENTS FOR RECTANGULAR WEIR WITH BOTTOM ORIFICES **15** 3265-80
[49] Farhan S L, Jasim I A and Al-Mamoori S K 2019 The transformation of the city of Najaf, Iraq: Analysis, reality and future prospects *Journal of Urban Regeneration & Renewal* **13** 160-71
[50] Hashim K S, Al Khaddar R, Jasim N, Shaw A, Phipps D, Kot P, Pedrola M O, Alattabi A W, Abdulredha M, Alawsh R J S and Technology P 2019 Electrocoagulation as a green technology for phosphate removal from River water **210** 135-44
[51] Hashim K S, Idowu I A, Jasim N, Al Khaddar R, Shaw A, Phipps D, Kot P, Pedrola M O, Alattabi A W and Abdulredha M J M 2018 Removal of phosphate from River water using a new baffle plates electrochemical reactor **5** 1413-8
[52] Majdi H S, Shubbar A, Nasr M S, Al-Khafaji Z S, Jafer H, Abdulredha M, Al Masoodi Z, Sadique M and Hashim K J D i B 2020 Experimental data on compressive strength and ultrasonic pulse velocity properties of sustainable mortar made with high content of GGBFS and CKD combinations
[53] Shubbar A A, Jafer H, Abdulredha M, Al-Khafaji Z S, Nasr M S, Al Masoodi Z and Sadique M J J o B E 2020 Properties of cement mortar incorporated high volume fraction of GGBFS and CKD from 1 day to 550 days **30** 101327

Removal of Pathogenic pollutants using electrocoagulation using aluminium electrodes

Shereen A Abdul-Husain[1], Sawsan Alramahi[2]
[1] Department of Civil Engineering, College of Engineering, University of Warith AL-Anbiya'a, Kerbala, Iraq.
Email: Shereen@uowast.edu.iq

Abstract. Pathogenic contamination of fresh waste bodies is some of the most harmful types of water contamination where in aquatic environment the pathogenic microorganisms can reproduce, which may pollute the marine environment overall. In addition, there are serious side effects with microbial infection. Some many pasteurization processes, like membrane separation and additives, have therefore been used to remove microorganisms from wastewater. Most of these processes are inefficient or hazardous, like the chemical treatment. Lately, electrocoagulation process, due to the environmental safety and comparatively low operational costs, has paid considerable attention to remove pathogenic contamination of sewage treatment plant effluent. During this analysis, the consequences of the changing of the electrical current change from 0.5 to 1.5 mA/cm^2 were examined on the ability of aluminium-based electrocoagulation unit for sewage treatment by disabling pathogenic pollutants. The findings demonstrated that the density of the current was increased to increasing the E. coli elimination (an indicator of pathogenic pollutants removal) to reach a current density of 2 mA/cm^2 the strongest deactivation (83.2%). Rising the current density increases the bacterial removal. However, the increase in the current density increases the treatment cost.

Keyword: Electrocoagulation, pathogenic contamination, aluminium.

1. Introduction

The population of the world is now subjected to various pollution sources, such as water, air, and soil pollution, which adversely affect human welling and life on earth [1, 2]. The contamination of the water, which can be described as the existence of undesirable contaminants in water, is an emerging burden due to wastewater disposal or the normal occurrence of a contaminant's overconcentration [3-5]. The contamination of the water is regard as one of the most harmful type of contamination due to the low volume of potable water on this planet which is less than 0.3 percent of earth water bodies and because the incapability of people to live without water for very short time [6-8]. In everyday lives, industrial sector and large-scale farming also contribute to a worrying trend towards contamination of water bodies both ground and surface water bodies [9-12]. One clear indication is a 75% rise in water bodies phosphate quantity compared with its pre-industrial revolution concentrations [13, 14].
Regrettably, like phosphorus, the levels of certain pollutants rose by so many times owing to industrialisation, include but are not restricted to nutrients [15-18], toxic substances [19-23], dyes [24, 25], fluoride [26-29], phenols [30], and organic material [31]. Water problems have been more severe because of the effect on water use of climate change[32-36], sharing on Earth planet of water and

people[37-41]. The oil industry which has been exponentially expanding over the last decade, for example, generates large quantities of wastewater containing high levels of toxic toxins, like arsenic, organic compounds, and biological contaminants. As described earlier, water contaminants, including toxic substances, phenols, painting chemicals, organic compounds and fertilizer, are sadly endless, leaving about half of the total worldwide population lacking healthy water supply in the upcoming years [42-45].

Biological contaminants are regarded as the most hazardous contaminants among water contaminates owing to their potential to substantial increase in numbers inside the water in comparatively short times and their serious effects on human health [22, 46]. Exceeding 1 million people have been confirmed to die worldwide annually due to diseases linked to water contamination. In addition, it has been identified that a significant number of water bodies are biologically contaminated in developed nations; for example, approximately 1/3 of Bangladesh's underground water sources were considered contaminated with biological pollutants. The lack of successful techniques of sterilization, in particular in developing countries, actually increased the biological contamination of water bodies [22]. Accordingly, many disinfection processes, including chlorine, UV radiation, magnetic charges and metallic salts, have been used to eliminate microbial contaminants from polluted water [21]. In a synthetic wastewater which includes toxic metals, nanoparticles of magnesium metal oxide have been used to disable the E. coli, and it was found that nanomaterials of magnesium metal ions completely kill all E. coli in 30 minutes treatment [47, 48]. Many technologies, though, are costly like nanoparticles or dangerous because they are producing toxic by-products like chlorination and chemical treatment [49]. Furthermore, certain technologies like biological reactors produces huge sludge volumes that involve costly dewatering processes before they are disposed of [50] or adopted as recycling materials [51-57].

In contrast to many other approaches, current advances of the electrocoagulation (EC) as a disinfection procedure for polluted wastewater have shown that biological contaminants can be removed within such a short amount of time [3, 28, 58]. The EC processes are also incredibly safe since there is no need of chemicals to be add to the water and harmful by-products from the reaction will not be generated [49]. To give a good instance, over 95 percent of Escherichia coli can be removed from of the artificial water sample in just about twenty minutes at a low operational cost, using Aluminium alloys EC units [49, 59]. In addition to the afore mentioned benefits, automating EC reactors with a suitable form of detectors like heat wave detectors is very simple [60-63]. Nevertheless, the effectiveness of the EC systems was also shown to rely on a number of operating factors like applied voltage, current density and adsorbent dosage [1, 16]. The effects of the density of the current on the removal of the E. coli's from municipal wastewater were investigated in this research.

2. Materials and methods

2.1. Experimental Setup

Experimentation was started using a 1500 mL (shape cylinder) plastic container, consisting of two aluminium electrodes. each electrode has a surface area of 450 cm2). Tests were initiated electrocoagulation device Because of its economy, large accessibility (including in underdeveloped nations), its low oxide potential, aluminium was included in this device [2]. Two holes were attached to the container, the high hole was used as an entrance for the liquid, the lower hole was used as outlet to extract samples to further analyse. To provide regulate the needed current density, the aluminium electrode was directly coupled a DC source of power.

The treatment using the EC was performed by the addition of 750 mL of the sample prepared within the device, and the supply voltage for the appropriate duration was turned on. All parameters of operation are maintained in all tests, except for the applied voltage. The experiments were conducted at 7.5 pH level (the true pH of the sewage specimen), time is 20 minutes and the temperature kept at 20 °C. The density of the electrical current density has been selected as a control variable in this analysis, as it

played a huge part in the Ec system. The density of the electrical current has been demonstrated to regulate the thermal degradation of electrode (generating metal oxides), which regulate the quality of elimination of selected contaminants [5]. Besides, cathodes are often used to produce Hydrogen ions that is accountable for raising the coagulated contaminants in foam onto the solution's surface [5].
The treatment method has been introduced at three separate current densities that are 0.5 mA/cm2, 1 mA/cm2 and 1.5 mA/cm2 to examine the impact of the current density on the elimination of the microbes from municipal sewage. The performance of neutralization of the bacteria using the EC technique was measured using the following equation [49]:

$$Neutralization(\%) = \frac{Start\ cell\ count\ \ -End\ cell\ count}{Start\ cell\ count} \times 100 \qquad (1)$$

The current density influence was measured based on the power usage of the aluminium EC using following equation:

$$Energy\ (kWh/m^3) = \frac{Current\ density\ (ampere) \times Cell\ potential\ (voltage\ ) \times\ treatment\ time\ (hour)}{1000 \times Volume\ of\ water\ sample\ (m^3)} \qquad (2)$$

2.2. Wastewater samples
During February 2020, a sewage samples were taken utilizing plastic bottles at Karbala wastewater treatment plant, situated in the city of Karbala, Iraq where each sample has a size of 3 litres capacity. The E. coli cells were used as indicator to investigate the performance of the EC system in terms of bacterial disinfection of municipal wastewater. The analytical procedure was undertaken based on the current methodology proposed by the American Public Health Association using the filtration membrane technique. After 24-hour incubation time at 35 °C, the number of E. coli colonies was recorded before and after the sewage treatment.

3. Results
The electrical current density regulates the removal of contaminants, as was discussed in earlier part of this research, through both regulation of the generated metal oxides and hydrogen ions within the EC reactor. The results of changing the current density of the neutralization of the E. coli have been examined using 0.5 and 1 and 1.5 mA/cm^2. The pH level of the sewerage was 7.5 and the treatment duration was 20 minutes. The temperature of the samples was 20 °C and the size was 1500 mL. The influence of the electrical current density on the neutralization of the E. coli is presented in Figure 1. The removal efficiency of E. coli was found to boost with elevated current densities. it was found that it was at a current density of 0.5 mA/cm^2, the removal percentage was about 35 percent and increased significantly to be around 83.5 percent using a current density of 1.5 mA/cm^2. As stated above, the change in the neutralization of E. coli at elevated current stems from a rise in the quantity of metal oxides produced and the production of the hydrogen ions, resulted in a significant enhancement in E. coli neutralization.

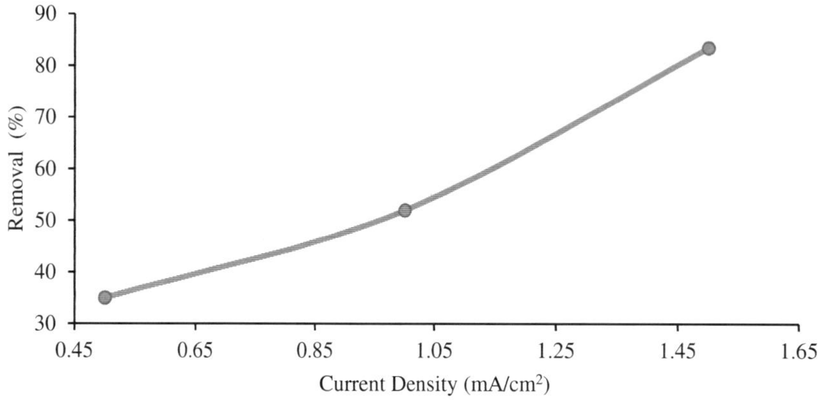

Figure 1: Electrical current density influence on E. coli removal.

Figure 2 shows the detrimental consequences of raising the current density. The raising current density leads to a significant rise in energy usage. The current density increased from 0.5 to 1.5 mA/cm^2 led to an increase the electricity usage from 2.1 to 6.3 kW/h/m^3. Because of these findings, while existing densities is highly useful for the elimination of biological contaminants from water, their performance must be improved such that high energy losses can be avoided. Furthermore, the efficiency of the aluminium EC have been noticed to be in reasonable accordance with the literature [49].

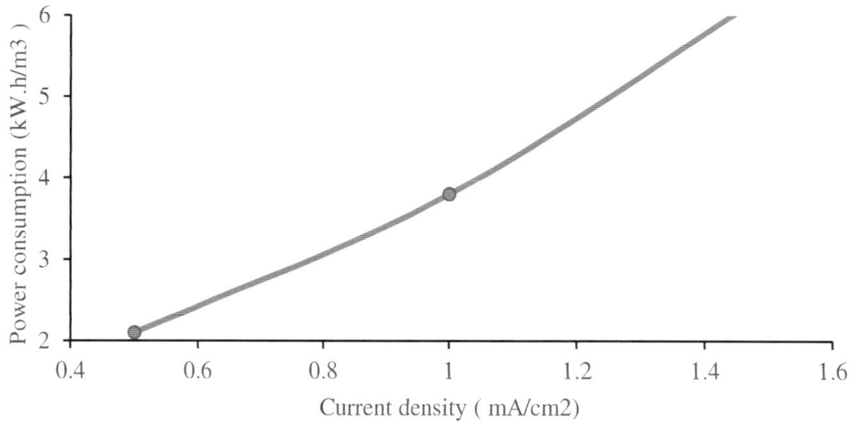

Figure 2: The power usage based on the current density supplied.

4. Conclusion

The present work was carried out to assess the impact on the neutralization of E. coli by the EC approach of the current density. The findings revealed that the current density should never be overlooked in the electrocoagulation removal method of microbial from swage, and the neutralization of E. coli improved approximately 50 percent by raising the current density from 0.5 to 1.5 mA/cm^2. This shows that the efficiency of aluminium EC can be enhanced by increasing the density of the electrical current. Nevertheless, the high-power usage increases energy demand, and hence the use of aluminium EC systems in the area of sewerage treatment may be limited. Additionally, parameters like electrode spacing and sewage temperature also play a significant role in the EC process, and further experiments must be conducted with the effects of the experimental variables on E. coli removal from sewage.

References

[1] Abdulhadi B A, Kot P, Hashim K S, Shaw A and Khaddar R A 2019 Influence of current density and electrodes spacing on reactive red 120 dye removal from dyed water using electrocoagulation/electroflotation (EC/EF) process. In: *First International Conference on Civil and Environmental Engineering Technologies (ICCEET),* (University of Kufa, Iraq pp 12-22

[2] Hashim K S, Al-Saati N H, Alquzweeni S S, Zubaidi S L, Kot P, Kraidi L, Hussein A H, Alkhaddar R, Shaw A and Alwash R 2019 Decolourization of dye solutions by electrocoagulation: an investigation of the effect of operational parameters. In: *First International Conference on Civil and Environmental Engineering Technologies (ICCEET),* (University of Kufa, Iraq pp 25-32

[3] Hashim K S, Idowu I A, Jasim N, Al Khaddar R, Shaw A, Phipps D, Kot P, Pedrola M O, Alattabi A W and Abdulredha M 2018 Removal of phosphate from River water using a new baffle plates electrochemical reactor *MethodsX* **5** 1413-8

[4] Hashim K S, Shaw A, Al Khaddar R, Ortoneda Pedrola M and Phipps D 2017 Defluoridation of drinking water using a new flow column-electrocoagulation reactor (FCER) - Experimental, statistical, and economic approach *Journal of Environmental Management* **197** 80-8

[5] Hashim K S, Shaw A, Al Khaddar R, Pedrola M O and Phipps D 2017 Iron removal, energy consumption and operating cost of electrocoagulation of drinking water using a new flow column reactor *Journal of Environmental Management* **189** 98-108

[6] Al-Marri S, AlQuzweeni S S, Hashim K S, AlKhaddar R, Kot P, AlKizwini R S, Zubaidi S L and Al-Khafaji Z S 2020 Ultrasonic-Electrocoagulation method for nitrate removal from water. In: *IOP Conference Series: Materials Science and Engineering*: IOP Publishing) p 012073

[7] Zanki A K, Mohammad F H, Hashim K S, Muradov M, Kot P, Kareem M M and Abdulhadi B 2020 Removal of organic matter from water using ultrasonic-assisted electrocoagulation method. In: *IOP Conference Series: Materials Science and Engineering*: IOP Publishing) p 012033

[8] Abdulhadi B, Kot P, Hashim K, Shaw A, Muradov M and Al-Khaddar R 2021 Continuous-flow electrocoagulation (EC) process for iron removal from water: Experimental, statistical and economic study *Science of The Total Environment* **756** 1-16

[9] Isra'a S S, Al-Janabi A, Abdulredha M, Alkandari A, Abdellatif M and Yeboah D 2021 Reusing of furnace bottom ash as an adsorbent for phosphate removal from water. In: *IOP Conference Series: Materials Science and Engineering*: IOP Publishing) p 012006

[10] Jawad S F, Saddam N S, Adaami Q J, Kareem M M, Abdulredha M, Mubarak H A, Kot P, Gkantou M and AlKhayyat A 2021 Dye removal from textile wastewater using solar-powered electrocoagulation reactor. In: *IOP Conference Series: Materials Science and Engineering*: IOP Publishing) p 012016

[11] Majdi H S, Shubbar A, Nasr M S, Al-Khafaji Z S, Jafer H, Abdulredha M, Al Masoodi Z, Sadique M and Hashim K J D i B 2020 Experimental data on compressive strength and ultrasonic pulse velocity properties of sustainable mortar made with high content of GGBFS and CKD combinations

[12] Shubbar A A, Jafer H, Abdulredha M, Al-Khafaji Z S, Nasr M S, Al Masoodi Z and Sadique M J J o B E 2020 Properties of cement mortar incorporated high volume fraction of GGBFS and CKD from 1 day to 550 days **30** 101327

[13] Alenezi A K, Hasan H A, Hashim K S, Amoako-Attah J, Gkantou M, Muradov M, Kot P and Abdulhadi B 2020 Zeolite-assisted electrocoagulation for remediation of phosphate from calcium-phosphate solution. In: *IOP Conference Series: Materials Science and Engineering*: IOP Publishing) p 012031

[14] Hashim K S, AlKhaddar R, Shaw A, Kot P, Al-Jumeily D, Alwash R and Aljefery M H 2020 *Electrocoagulation as an eco-friendly River water treatment method. In Advances in Water Resources Engineering and Management* (Berline: Springer)

[15] Alenazi M, Hashim K S, Hassan A A, Muradov M, Kot P and Abdulhadi B 2020 Turbidity removal using natural coagulants derived from the seeds of strychnos potatorum: statistical and experimental approach. In: *IOP Conference Series: Materials Science and Engineering*: IOP Publishing) p 012064

[16] Mohammed A-H, Hussein A H, Yeboah D, Al Khaddar R, Abdulhadi B, Shubbar A A and Hashim K S 2020 Electrochemical removal of nitrate from wastewater. In: *IOP Conference Series: Materials Science and Engineering*: IOP Publishing) p 012037

[17] Alattabi A W, Harris C, Alkhaddar R, Alzeyadi A and Hashim K 2017 Treatment of Residential Complexes' Wastewater using Environmentally Friendly Technology *Procedia Engineering* **196** 792-9

[18] Alattabi A W, Harris C B, Alkhaddar R M, Hashim K S, Ortoneda-Pedrola M and Phipps D 2017 Improving sludge settleability by introducing an innovative, two-stage settling sequencing batch reactor *Journal of Water Process Engineering* **20** 207-16

[19] Abdulla G, Kareem M M, Hashim K S, Muradov M, Kot P, Mubarak H A, Abdellatif M and Abdulhadi B 2020 Removal of iron from wastewater using a hybrid filter. In: *IOP Conference Series: Materials Science and Engineering*: IOP Publishing) p 012035

[20] Abdulraheem F S, Al-Khafaji Z S, Hashim K S, Muradov M, Kot P and Shubbar A A 2020 Natural filtration unit for removal of heavy metals from water. In: *IOP Conference Series: Materials Science and Engineering*: IOP Publishing) p 012034

[21] Hashim K S, Ali S S M, AlRifaie J K, Kot P, Shaw A, Al Khaddar R, Idowu I and Gkantou M 2020 Escherichia coli inactivation using a hybrid ultrasonic–electrocoagulation reactor *Chemosphere* **247** 125868-75

[22] Hashim K S, Shaw A, AlKhaddar R, Kot P and Al-Shamma'a A 2021 Water purification from metal ions in the presence of organic matter using electromagnetic radiation-assisted treatment *Journal of Cleaner Production* **280**

[23] Hashim K S, Ewadh H M, Muhsin A A, Zubaidi S L, Kot P, Muradov M, Aljefery M and Al-Khaddar R 2020 Phosphate removal from water using bottom ash: Adsorption performance, coexisting anions and modelling studies *Water Science and Technology* **83** 1-17

[24] Aqeel K, Mubarak H A, Amoako-Attah J, Abdul-Rahaim L A, Al Khaddar R, Abdellatif M, Al-Janabi A and Hashim K S 2020 Electrochemical removal of brilliant green dye from wastewater. In: *IOP Conference Series: Materials Science and Engineering*: IOP Publishing) p 012036

[25] Hashim K S, Hussein A H, Zubaidi S L, Kot P, Kraidi L, Alkhaddar R, Shaw A and Alwash R 2019 Effect of initial pH value on the removal of reactive black dye from water by electrocoagulation (EC) method. In: *2nd International Scientific Conference,* (Al-Qadisiyah University, Iraq pp 12-22

[26] Alhendal M, Nasir M J, Hashim K S, Amoako-Attah J, Al-Faluji D, Muradov M, Kot P and Abdulhadi B 2020 Cost-effective hybrid filter for remediation of water from fluoride. In: *IOP Conference Series: Materials Science and Engineering*: IOP Publishing) p 012038

[27] Alyafei A, AlKizwini R S, Hashim K S, Yeboah D, Gkantou M, Al Khaddar R, Al-Faluji D and Zubaidi S L 2020 Treatment of effluents of construction industry using a combined filtration-electrocoagulation method. In: *IOP Conference Series: Materials Science and Engineering*: IOP Publishing) p 012032

[28] Hashim K S, Khaddar R A, Jasim N, Shaw A, Phipps D, Kot P, Pedrola M O, Alattabi A W, Abdulredha M and Alawsh R 2019 Electrocoagulation as a green technology for phosphate removal from River water *Separation and Purification Technology* **210** 135-44

[29] Hassan Alnaimi I J I, Abuduljaleel Al-Janabi, Khalid Hashim, Michaela Gkantou, Salah L. Zubaidi, Patryk Kot, Magomed Muradov 2020 Ultrasonic-electrochemical treatment for effluents of

concrete plants Ultrasonic-electrochemical treatment for effluents of concrete plants. In: *IOP Conference Series Materials Science and Engineering* (University of Kufa, Najaf, Iraq pp 1-9

[30] Emamjomeh M M, Mousazadeh M, Mokhtari N, Jamali H A, Makkiabadi M, Naghdali Z, Hashim K S and Ghanbari R 2020 Simultaneous removal of phenol and linear alkylbenzene sulfonate from automotive service station wastewater: Optimization of coupled electrochemical and physical processes *Separation Science and Technology* **55** 3184-94

[31] Al-Saati N H, Hussein T K, Abbas M H, Hashim K, Al-Saati Z N, Kot P, Sadique M, Aljefery M H and Carnacina I 2019 Statistical modelling of turbidity removal applied to non-toxic natural coagulants in water treatment: a case study *Desalination and Water Treatment* **150** 406-12

[32] Alattabi A W, Harris C, Alkhaddar R, Alzeyadi A and Abdulredha M J P e 2017 Online monitoring of a sequencing batch reactor treating domestic wastewater **196** 800-7

[33] Al-Sareji O J, Abdulredha M, Mubarak H A, Grmasha R A, Alnowaishry A, Kot P, Al-Khaddar R and AlKhayyat A 2021 Copper removal from water using carbonized sawdust. In: *IOP Conference Series: Materials Science and Engineering*: IOP Publishing) p 012015

[34] ALWAN H H, SALEH L A, AL-MOHAMMED F M, ABDULREDHA M A J J o E S and Technology 2020 EXPERIMENTAL PREDICTION OF THE DISCHARGE COEFFICIENTS FOR RECTANGULAR WEIR WITH BOTTOM ORIFICES **15** 3265-80

[35] Hashim K S, Al Khaddar R, Jasim N, Shaw A, Phipps D, Kot P, Pedrola M O, Alattabi A W, Abdulredha M, Alawsh R J S and Technology P 2019 Electrocoagulation as a green technology for phosphate removal from River water **210** 135-44

[36] Hashim K S, Idowu I A, Jasim N, Al Khaddar R, Shaw A, Phipps D, Kot P, Pedrola M O, Alattabi A W and Abdulredha M J M 2018 Removal of phosphate from River water using a new baffle plates electrochemical reactor **5** 1413-8

[37] Abdulredha M, Kot P, Al Khaddar R, Jordan D, Abdulridha A J E, Development and Sustainability 2020 Investigating municipal solid waste management system performance during the Arba'een event in the city of Kerbala, Iraq **22** 1431-54

[38] Abdulredha M, Muhsin A A, Al-Janabi A, Alajmi B N, Gkantou M, Amoako-Attah J, Al-Jumeily D, Mustafina J and AlKhayyat A 2021 Using SF and CKD as cement replacement materials for producing cement mortar. In: *IOP Conference Series: Materials Science and Engineering*: IOP Publishing) p 012007

[39] Abdulredha M, Rafid A, Jordan D and Alattabi A J P e 2017 Facing up to waste: how can hotel managers in Kerbala, Iraq, help the city deal with its waste problem? **196** 771-8

[40] Abdulredha M, Rafid A, Jordan D and Hashim K J P E 2017 The development of a waste management system in Kerbala during major pilgrimage events: determination of solid waste composition **196** 779-84

[41] Al-Anbari R, Alnakeeb A, Abdulredha M J E and Journal T 2013 Landfill site selection for Kerbala municipal solid wastes by using geographical information system techniques **32** 13

[42] Abdulredha M, Al Khaddar R, Jordan D, Al-Attabi A and Alzeyadi A 2017 Public participation in solid waste management during mega festivals: A pilot study. In: *WCST World Congress on Sustainable Technologies Proceedings 2017*: Infonomics Society) pp 38-41

[43] Abdulredha M, Al Khaddar R, Jordan D, Kot P, Abdulridha A and Hashim K J W M 2018 Estimating solid waste generation by hospitality industry during major festivals: A quantification model based on multiple regression **77** 388-400

[44] Abdulredha M, al-Khaddar R, Kot P, Jordan D and Abdulridha A 2018 Benchmarking of the Current Solid Waste Management System in Karbala, Iraq, Using Wasteaware Benchmark Indicators. In: *World Environmental and Water Resources Congress 2018: Groundwater, Sustainability, and Hydro-Climate/Climate Change*: American Society of Civil Engineers Reston, VA) pp 40-8

[45] Abdulredha M, Kadhim N, Hussein A, Almutairi M, Alkhaddar R, Yeboah D and Hashim K 2021 Zeolite as a natural adsorbent for nitrogenous compounds being removed from water. In: *IOP Conference Series: Materials Science and Engineering*: IOP Publishing) p 012082
[46] Hashim K S, Al-Saati N H, Hussein A H and Al-Saati Z N 2018 An investigation into the level of heavy metals leaching from canal-dreged sediment: a case study metals leaching from dreged sediment. In: *First International Conference on Materials Engineering & Science,* (Istanbul Aydın University (IAU), Turkey pp 12-22
[47] Cai Y, Li C, Wu D, Wang W, Tan F, Wang X, Wong P K and Qiao X 2017 Highly active MgO nanoparticles for simultaneous bacterial inactivation and heavy metal removal from aqueous solution *Chemical Engineering Journal* **312** 158-66
[48] Abdulredha M, Abdulridha A, Shubbar A, Alkhaddar R, Kot P and Jordan D 2020 Estimating municipal solid waste generation from service processions during the Ashura religious event. In: *IOP Conference Series: Materials Science and Engineering*: IOP Publishing) p 012075
[49] Hashim K, Kot P, Zubaid S, Alwash R, Al Khaddar R, Shaw A, Al-Jumeily D and Aljefery M 2020 Energy efficient electrocoagulation using baffle-plates electrodes for efficient Escherichia Coli removal from Wastewater *Journal of Water Process Engineering* **33** 101079-86
[50] Idowu I A, Atherton W, Hashim K, Kot P, Alkhaddar R, Alo B I and Shaw A 2019 An analyses of the status of landfill classification systems in developing countries: Sub Saharan Africa landfill experiences *Waste Management* **87** 761-71
[51] Shubbar A A, Jafer H, Dulaimi A, Hashim K, Atherton W and Sadique M 2018 The development of a low carbon binder produced from the ternary blending of cement, ground granulated blast furnace slag and high calcium fly ash: An experimental and statistical approach *Construction and Building Materials* **187** 1051-60
[52] Shubbar A A, Al-Shaer A, AlKizwini R S, Hashim K, Hawesah H A and Sadique M 2019 Investigating the influence of cement replacement by high volume of GGBS and PFA on the mechanical performance of cement mortar. In: *First International Conference on Civil and Environmental Engineering Technologies (ICCEET),* (University of Kufa, Iraq pp 31-8
[53] Kadhim A, Sadique M, Al-Mufti R and Hashim K 2020 Long-term performance of novel high-calcium one-part alkali-activated cement developed from thermally activated lime kiln dust *Journal of Building Engineering* **32** 1-17
[54] Kadhim A, Sadique M, Al-Mufti R and Hashim K 2020 Developing One-Part Alkali-Activated metakaolin/natural pozzolan Binders using Lime Waste as activation Agent *Advances in Cement Research* **32** 1-38
[55] Majdi H S, Shubbar A, Nasr M S, Al-Khafaji Z S, Jafer H, Abdulredha M, Masoodi Z A, Sadique M and Hashim K 2020 Experimental data on compressive strength and ultrasonic pulse velocity properties of sustainable mortar made with high content of GGBFS and CKD combinations *Data in Brief* **31** 105961-72
[56] Shubbar A A, Sadique M, Nasr M S, Al-Khafaji Z S and Hashim K S 2020 The impact of grinding time on properties of cement mortar incorporated high volume waste paper sludge ash *Karbala International Journal of Modern Science* **6** 1-23
[57] Shubbar A A, Sadique M, Shanbara H K and Hashim K 2020 *The Development of a New Low Carbon Binder for Construction as an Alternative to Cement. In Advances in Sustainable Construction Materials and Geotechnical Engineering* (Berlin: Springer)
[58] Aayef A N, Al Masoodi W T M, Kamel R J, Abdulredha M, Almansoori N A, Kot P and Muradov M 2021 An experimental study for adapting electrocoagulation as a technique for fluoride removal from water. In: *IOP Conference Series: Materials Science and Engineering*: IOP Publishing) p 012012

[59] Abdulredha M, Al Khaddar R and Jordan D 2017 Hoteliers' attitude towards solid waste source separation through mega festivals: A pilot study in Karbala. In: *International Conference for Doctoral Research*: BUID)

[60] Gkantou M, Muradov M, Kamaris G S, Hashim K, Atherton W and Kot P 2019 Novel Electromagnetic Sensors Embedded in Reinforced Concrete Beams for Crack Detection *Sensors* **19** 5175-89

[61] Ryecroft S, Shaw A, Fergus P, Kot P, Hashim K, Moody A and Conway L 2019 A First Implementation of Underwater Communications in Raw Water Using the 433 MHz Frequency Combined with a Bowtie Antenna *Sensors* **19** 1813-23

[62] Ryecroft S P, shaw A, Fergus P, Kot P, Hashim K and Conway L 2019 A Novel Gesomin Detection Method Based on Microwave Spectroscopy. In: *12th International Conference on Developments in eSystems Engineering (DeSE),* (Kazan, Russia pp 429-33

[63] Teng K H, Kot P, Muradov M, Shaw A, Hashim K, Gkantou M and Al-Shamma'a A 2019 Embedded Smart Antenna for Non-Destructive Testing and Evaluation (NDT&E) of Moisture Content and Deterioration in Concrete *Sensors* **19** 547-59

Using Partial Substitution of Portland Cement with Silica Fume and CKD to Determine the Ultrasonic Pulse Velocity of Cement Mortar

Waeel A. Radhi[1]
[1] Faculty of Engineering, Basrah University, Basarh, Iraq

Email: radhi-92@yahoo.com

Abstract. Portland cement considered one of the most pollutant material used in the construction industry, which is responsible for several harmful emissions like a greenhouse gas. This causes particular environmental interest in the production of concrete and its components. for that, it becomes necessary to find an alternative material for ordinary Portland cement with good environmental impact as well as good construction properties. The aim of this paper is to investigate the effects of partial replacement of ordinary Portland cement by industrial by-products substances in the cement mortar. The industrial by-products used in the present research is silica fume and cement kiln dust (CKD). Various mortar specimens are prepared by replacing the in different proportion of 20%, 40, and 60% of a mixture from CKD and SF. the testing of samples were conducted after curing the specimens for 7-days and 28-days. The strength of specimens was tested by ultrasonic pulse velocity (UPV) test. The results show that the optimum percentage of replacement Portland cement by silica fume and CKD was involved from 20% to 28%. The increment in the percentage of CKD& silica fume till the optimum percentage shows good enhancement in the pulse velocity of samples, while the additional quantities of these by-products start to reduce the pulse velocity of samples.

Keywords: ultrasonic pulse velocity, cement kiln dust, silica fume, E. coli, wastewater.

1. Introduction

Concrete is a critical component of today's building sector, and Portland Cement has to be the most significant element in the mixture [1]. Although cement's critical position in the maintenance of society and the world economy, the cement function in environmental degradation is well-established [2]. According to estimates, cement plants release (CO_2) equal to 6 to 8% of the total (CO_2) in the atmosphere [3, 4], resulting in tragic environmental consequences also including climate change[5-7]. as a result of the above, the wastewater emissions [8-10] and water intake have increased dramatically [11-14]. Furthermore, owing to the chemical structure of Portland cement, discharged wastewaters from factories of concrete and pouring operations are quite specific (pH of 9 to 11) [15, 16]. resulting in tragic effects on aquatic life quality and causing the deterioration of the living organisms [17-19]. To extract several of the contaminants present in cement plant wastewater, like phenols [20-22], nitrates [23-26], metals [27-31], phosphate [23, 32-37] and turbidity [38-41], multiple wastewater treatments techniques were created [42-44]. Nonetheless, these strategies are insufficient to monitor the entire predicted emissions from cement plants [40, 43, 45-48].

Cement manufacturing has been an increasing source of concern considering the large amounts of (CO_2) emitted from cement-related activities to the atmosphere and pollutants released into the aquatic life. As a result, previous studies have concluded that the invention of substitutes that substitute as a component of cement, like silica fume and cement kiln dust (CKD), is the most practical alternatives for portland cement manufacturing [1, 49-51].

The use of silica fume and CKD has been shown to have negative impacts on concrete properties; however, these substances can boost the efficiency of the concrete. The flowability of fresh blends decreased as the ratio of CKD to cement raised. moreover, as maximising the CKD percentage, the concrete strength decreases [1]. Because of their large surface areas and consists of a significant amount of silica oxides, pozzolanic compounds, such as ultrafine silica fume, are very effective and commonly used as cementitious substances [16]. Previous studies showed that (0.22 - .30) is a good percentage to replace Portland cement with silica fume without a decrease in strength [4].

one of the non-destructive testing techniques is The ultrasonic pulse velocity (UPV), which utilized to determine the consistency of structures made of concrete. Some properties of concrete such as quality and strength are assessed using this process, which involves measuring the velocity of the ultrasonic pulse velocity passing through members of concretes [16, 52].

The efficiency of silica fume and CKD as a substitute for Portland cement was evaluated in this report. The key aim of this research is to use an ultrasonic pulse velocity test to check if these substances affect the cement properties such as durability at various curing times (7 and 28 days).

2. Curriculum of experimentation

The ultrasonic pulse velocity of the cement mixture created by partially replacing Portland cement with silica fume and CKD was determined in a series of laboratory experiments. Later on, the mixed pattern, as well as the percentage distribution of silica fume and CKD utilized in this survey, were dismissed.

2.1. Materials

The cement kiln dust (CKD) is a waste material generated by cement manufacturing, which is disposed of in landfills. The key properties of the CKD, which is a fine-grained, tough, highly alkaline powder substance. The raw materials, overall machinery configuration, kiln pyro-processing form, fuels, and type of cement being produced all influence these properties. CKD consist of different components such as limestone residuals, halide, alkali sulphates, as well as calcined and raw feed and other combustion byproducts. this material can be utilized as cementitious material with Ground Granulated Blast Slag and fly ash in different percentage up to 15%. Due to the high alkali content, if cement kiln dust is used individually, the resulting mixture may have reduced setting time, weight and workability and weight [53].

silica fume is another by-product that commonly used as cementitious materials in concrete due to its characteristics which improve cementitious reaction[54]. this product considered a secondary product of silicon and ferrosilicon alloy manufacturing industries[54]. generally, when silica fume applied in concrete, the concrete's strength and consistency are significantly improved. Instead of being disposed of in landfill site, the ashes released by furnaces are deposited as silica fume in the manufacturing of silicon alloys.

the partials of silica fume are very small and circular and about 100 times smaller than the normal cement particles.[54]. Because of the silica oxide that contributes to its structure, as well as its surface area and micro size, silica fume is considered a highly reactive substance.

Portland Cement was used as the major binding material in this experiment due to its excellent mechanical properties that serve the mixture to be coherent. BS EN 196-2:2013 was used to measure the cement used in this project.

The chemical structure of silica fume, (CKD) and Portland cement is seen in Figures 1,2, and 3. These characteristics comply with the BS-EN-197-1(2011) and BS-EN-450-1 standards (2012).

BS EN 12620:2002+A1 standard (2008) was utilized in this study to check the size distributions of the grains, as well as the chloride and sulphate contents.

clean river sands were utilized as fine aggregates in the concrete mixture in this analysis. Concrete was also mixed and cured with portable water that was free of Impurities and organic matter.

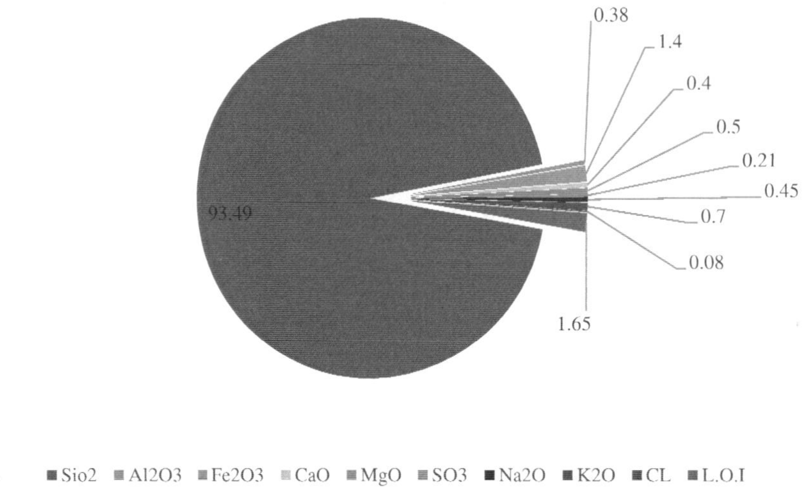

Figure (1): The chemical structure of silica fume

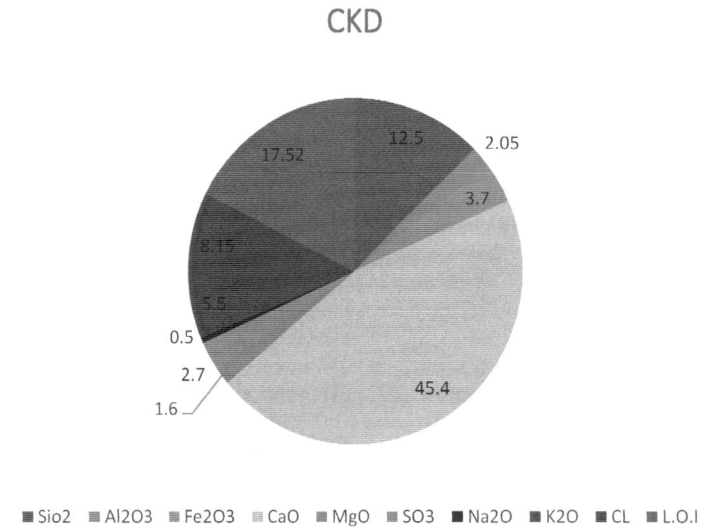

Figure (2): The chemical structure of CKD

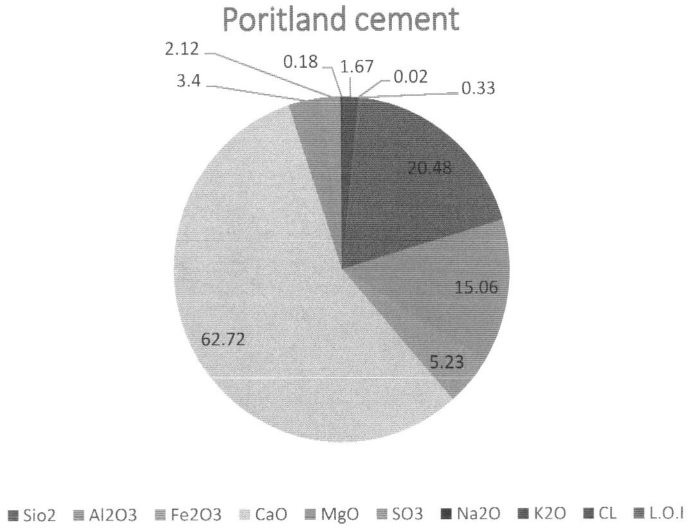

Figure (3): The chemical structure of Portland cement

2.2. Methods of testing

Three prisms (160x40x40 mm) were cast for each case to investigate the impact of partially replacing Portland cement with silica fume and CKD on the consistency of the mortar mixture. These experiments involve only analyzing materials pouring from cement mortar. The examined specimens are then made up of cement in three different silica fume and CKD concentrations. Both specimens were preserved in good shape and moulded and put in water for curing after one day of casting. Ultrasonic experiments were then performed on hardening samples at 7 and 28 days in accordance with BS EN 12504-4:2004.

2.3. Mix Design

casting the control cement mortars by selecting the sum of cement, water, fine aggregate, and mineral admixture materials is part of the construction process in this experimental investigation. in the design of mixtures, fine aggregates were chosen to Follow standard rating curves. In all of the blends, the water to (cement, silica fume and CKD) ratio is 0.41. One part cement to 2.5 parts sand was the amount of sand to a binder. Figure (4) shows the weight of each component/ingredient as well as the blend style percentage.

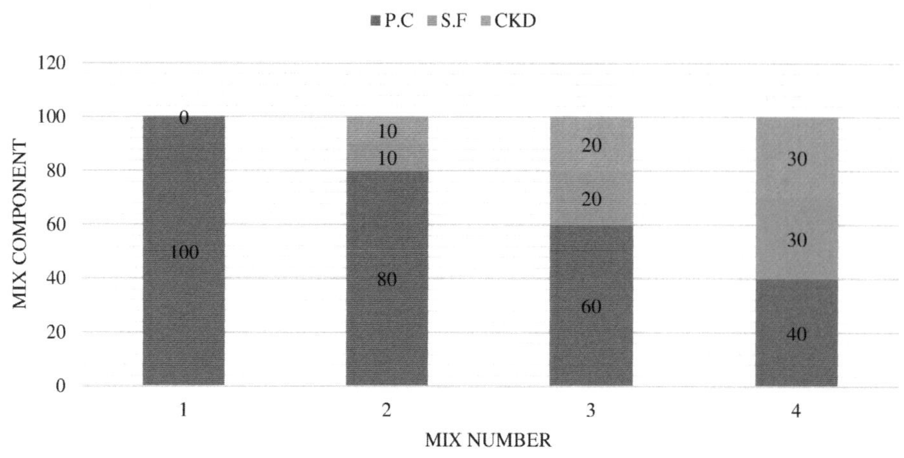

Figure (4): The Mix design.

UPV test is commenced by emitting ultrasonic pulses via the sample to be examined and calculating the required time by the pulse to penetrate the sample (direct method). High velocity indicates the good quality of the concrete sample, while slow velocity indicates the bad quality of the concrete sample. UPV testing equipment/tools include pulses generators, a transducer for the transformation of the electronic pulses into mechanical pulses having oscillations from 40 to 50 kHz, and a pulses receiver. Pulse's velocities are calculated as follows:

Pulse producers, a transducer for converting electrical signals into mechanical signals with Average vibrations ranging from 40 to 50 kilohertz, and a pulses detector are all used in UPV research. The following formula is used to measure pulse velocities:

$$\text{Pulse velosity} = \frac{\text{The specimen's thickness}}{\text{The required time for the pulse to penetrate the sample}} \quad (1)$$

3. Results

Table 1 and Figures 5 and 6 explain the effects of testing of a control cement mortar with partial substitution of Portland cement with varying amounts of silica fume and CKD at various curing times.

Table 1. Ultrasonic Pulse Velocity tests at 7-days and 28-days.

experiment No.	Portland cement (%)	Silica fume (%)	CKD (%)	UPV test (m/s)	
				7 days	28 days
1	100	0	0	3805	3965
2	80	10	10	3913	4064
3	60	20	20	3885	3982
4	40	30	30	3304	3647

The following can be deduced from these study results:

(i): It has been shown that using a fractional substitution of silica fume and CKD in mixtures reduces mortar pulse velocity values by a minor amount. anyway, as opposed to others, like the first mix, the second mix indicates an improvement in pulse velocity. For illustration, after 28 days of curing, the pulse velocity values of the three blends had improved by 2.5%, 0.5%, and -8%, respectively, in contrast to the first blend. Figures 5 and 6 show these results. This may be demonstrated by the fact that a limited quantity of cementitious materials will fill empty spaces in the mortar, which enhancing its adhesive characteristics.

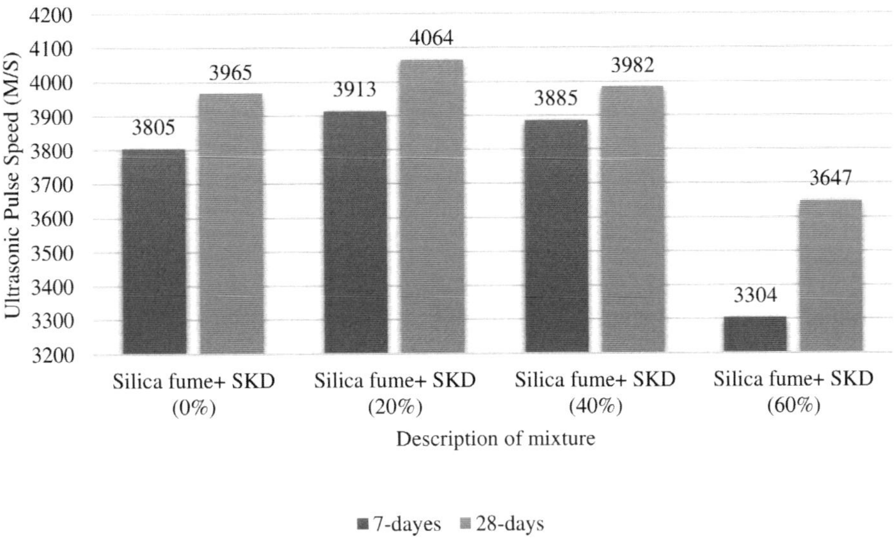

Figure 5. Ultrasonic Pulse Velocity Monitor for 7 and 28 curing day mixes.

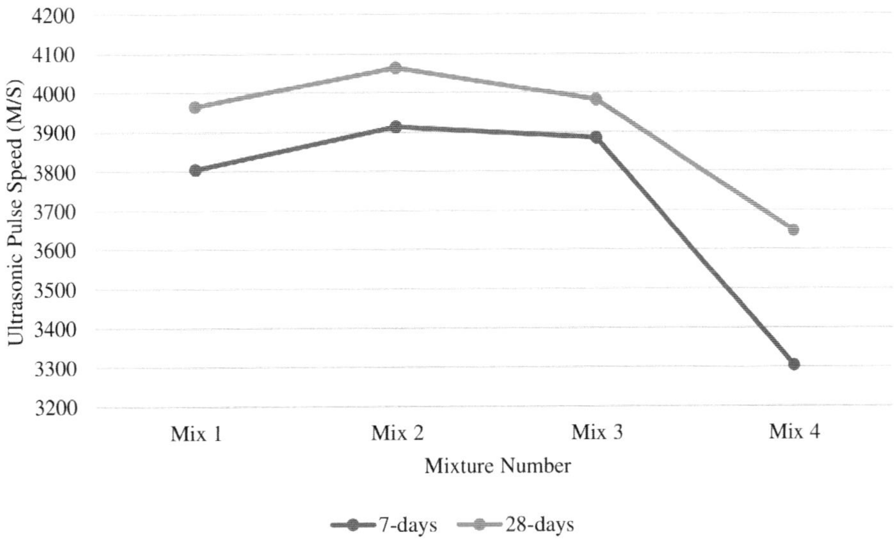

Figure 6. Impact of limited Silica fume and CKD substitution on mix longevity.

Utilizing 60% of a partial replacement material reduces the pulse velocity values of the mix by about 13% after 7 days of curing. since the silica fume and CKD are ineffective substance at the initial times of their use and need a period to associate with the compounds of cement, this value dropped to 8%

after 28 days of curing. In conclusion, partial substitution of silica fume and CKD should be achieved between 20% and 28% to achieve reasonable longevity and minimize emissions. This is based on the fact that supplementary cementitious materials decrease concrete compressive strength, which is an important element in the manufacturing gel (C-S-H) in concrete. For this, these react after solidification and use the moisture ingredients, Ca $(OH)_2$, to enable and begin silica fume and cement kiln dust hydration.

(ii): The findings suggest that samples that have been curing for 28 days have higher pulse velocity values. This is attributable to the fact that the curing period causes the benefit and improvement of C-S-H, which contributes to a reduction in the number of internal spaces or permeability in the concrete mixture, which influences the characteristics of concrete where increases the ability of concrete to resist the compressive forces.

The Ultrasonic Pulse velocity approach was used in this analysis, which can be classified as a conventional non-destructive test technique; thus, more modern techniques are available to verify the characteristics of concrete. In this sense, detectors have been used in earlier researches to check for concrete cracks [55], concrete humidity [56], and other implementations [57, 58]. Future research may use the same techniques.

4. Conclusions

Based on the findings of this research study it can be concluded that as Portland cement is partially replaced with silica fume and CKD in concrete, the pulse velocity values minimized as the proportion is increased. The quality of materials, on the other hand, indicates a modest increase with limited replacement levels. Whenever the cement is replaced by supplementary content in the mixture, a longer curing time results in a sample of better quality. Utilizing 20–28% of supplementary cementitious material as a partial substitute for Portland cement may be suitable ratios, with an increase in this ratio resulting in a small improvement in the sample quality. Furthermore, the applicable approach in this study was the Ultrasonic Pulse velocity, which can be classified as a conventional tool; thus, more modern techniques, such as Laser Scanning, are suggested to verify the mechanical characteristics of concrete.

References

[1] Shubbar A A, Jafer H, Dulaimi A, Hashim K, Atherton W and Sadique M 2018 The development of a low carbon binder produced from the ternary blending of cement, ground granulated blast furnace slag and high calcium fly ash: An experimental and statistical approach *Construction and Building Materials* **187** 1051-60

[2] Shubbar A A, Al-Shaer A, AlKizwini R S, Hashim K, Hawesah H A and Sadique M 2019 Investigating the influence of cement replacement by high volume of GGBS and PFA on the mechanical performance of cement mortar. In: *First International Conference on Civil and Environmental Engineering Technologies (ICCEET),* (University of Kufa, Iraq pp 31-8

[3] Kadhim A, Sadique M, Al-Mufti R and Hashim K 2020 Developing One-Part Alkali-Activated metakaolin/natural pozzolan Binders using Lime Waste as activation Agent *Advances in Cement Research* **32** 1-38

[4] Shubbar A A, Sadique M, Nasr M S, Al-Khafaji Z S and Hashim K S 2020 The impact of grinding time on properties of cement mortar incorporated high volume waste paper sludge ash *Karbala International Journal of Modern Science* **6** 1-23

[5] Aayef A N, Al Masoodi W T M, Kamel R J, Abdulredha M, Almansoori N A, Kot P and Muradov M 2021 An experimental study for adapting electrocoagulation as a technique for fluoride removal from water. In: *IOP Conference Series: Materials Science and Engineering*: IOP Publishing) p 012012

[6] Abdulredha M, Abdulridha A, Shubbar A, Alkhaddar R, Kot P and Jordan D 2020 Estimating municipal solid waste generation from service processions during the Ashura religious event. In: *IOP Conference Series: Materials Science and Engineering*: IOP Publishing) p 012075

[7] Abdulredha M, Al Khaddar R and Jordan D 2017 Hoteliers' attitude towards solid waste source separation through mega festivals: A pilot study in Karbala. In: *International Conference for Doctoral Research*: BUID)

[8] Abdulredha M, Al Khaddar R, Jordan D, Al-Attabi A and Alzeyadi A 2017 Public participation in solid waste management during mega festivals: A pilot study. In: *WCST World Congress on Sustainable Technologies Proceedings 2017*: Infonomics Society) pp 38-41

[9] Abdulredha M, Al Khaddar R, Jordan D, Kot P, Abdulridha A and Hashim K J W M 2018 Estimating solid waste generation by hospitality industry during major festivals: A quantification model based on multiple regression **77** 388-400

[10] Abdulredha M, al-Khaddar R, Kot P, Jordan D and Abdulridha A 2018 Benchmarking of the Current Solid Waste Management System in Karbala, Iraq, Using Wasteaware Benchmark Indicators. In: *World Environmental and Water Resources Congress 2018: Groundwater, Sustainability, and Hydro-Climate/Climate Change*: American Society of Civil Engineers Reston, VA) pp 40-8

[11] Abdulredha M, Kadhim N, Hussein A, Almutairi M, Alkhaddar R, Yeboah D and Hashim K 2021 Zeolite as a natural adsorbent for nitrogenous compounds being removed from water. In: *IOP Conference Series: Materials Science and Engineering*: IOP Publishing) p 012082

[12] Abdulredha M, Kot P, Al Khaddar R, Jordan D, Abdulridha A J E, Development and Sustainability 2020 Investigating municipal solid waste management system performance during the Arba'een event in the city of Kerbala, Iraq **22** 1431-54

[13] Abdulredha M, Muhsin A A, Al-Janabi A, Alajmi B N, Gkantou M, Amoako-Attah J, Al-Jumeily D, Mustafina J and AlKhayyat A 2021 Using SF and CKD as cement replacement materials for producing cement mortar. In: *IOP Conference Series: Materials Science and Engineering*: IOP Publishing) p 012007

[14] Abdulredha M, Rafid A, Jordan D and Alattabi A J P e 2017 Facing up to waste: how can hotel managers in Kerbala, Iraq, help the city deal with its waste problem? **196** 771-8

[15] Majdi H S, Shubbar A, Nasr M S, Al-Khafaji Z S, Jafer H, Abdulredha M, Masoodi Z A, Sadique M and Hashim K 2020 Experimental data on compressive strength and ultrasonic pulse velocity properties of sustainable mortar made with high content of GGBFS and CKD combinations *Data in Brief* **31** 105961-72

[16] Shubbar A A, Sadique M, Shanbara H K and Hashim K 2020 *The Development of a New Low Carbon Binder for Construction as an Alternative to Cement. In Advances in Sustainable Construction Materials and Geotechnical Engineering* (Berlin: Springer)

[17] Alyafei A, AlKizwini R S, Hashim K S, Yeboah D, Gkantou M, Al Khaddar R, Al-Faluji D and Zubaidi S L 2020 Treatment of effluents of construction industry using a combined filtration-electrocoagulation method. In: *IOP Conference Series: Materials Science and Engineering*: IOP Publishing) p 012032

[18] Abdulraheem F S, Al-Khafaji Z S, Hashim K S, Muradov M, Kot P and Shubbar A A 2020 Natural filtration unit for removal of heavy metals from water. In: *IOP Conference Series: Materials Science and Engineering*: IOP Publishing) p 012034

[19] Abdulla G, Kareem M M, Hashim K S, Muradov M, Kot P, Mubarak H A, Abdellatif M and Abdulhadi B 2020 Removal of iron from wastewater using a hybrid filter. In: *IOP Conference Series: Materials Science and Engineering*: IOP Publishing) p 012035

[20] Aqeel K, Mubarak H A, Amoako-Attah J, Abdul-Rahaim L A, Al Khaddar R, Abdellatif M, Al-Janabi A and Hashim K S 2020 Electrochemical removal of brilliant green dye from wastewater. In: *IOP Conference Series: Materials Science and Engineering*: IOP Publishing) p 012036

[21] Emamjomeh M M, Mousazadeh M, Mokhtari N, Jamali H A, Makkiabadi M, Naghdali Z, Hashim K S and Ghanbari R 2020 Simultaneous removal of phenol and linear alkylbenzene sulfonate from automotive service station wastewater: Optimization of coupled electrochemical and physical processes *Separation Science and Technology* **55** 3184-94

[22] Hashim K, Shaw A and Phipps D 2019 Treatment reactor and method of treating a liquid. In: *WIPO,* ed L J M University (United Kingdom

[23] Hashim K S, Ali S S M, AlRifaie J K, Kot P, Shaw A, Al Khaddar R, Idowu I and Gkantou M 2020 Escherichia coli inactivation using a hybrid ultrasonic–electrocoagulation reactor *Chemosphere* **247** 125868-75
[24] Hashim K S, AlKhaddar R, Shaw A, Kot P, Al-Jumeily D, Alwash R and Aljefery M H 2020 *Electrocoagulation as an eco-friendly River water treatment method. In Advances in Water Resources Engineering and Management* (Berline: Springer)
[25] Hassan Alnaimi I J I, Abuduljaleel Al-Janabi, Khalid Hashim, Michaela Gkantou, Salah L. Zubaidi, Patryk Kot, Magomed Muradov 2020 Ultrasonic-electrochemical treatment for effluents of concrete plants Ultrasonic-electrochemical treatment for effluents of concrete plants. In: *IOP Conference Series Materials Science and Engineering* (University of Kufa, Najaf, Iraq pp 1-9
[26] Hashim K S, Khaddar R A, Jasim N, Shaw A, Phipps D, Kot P, Pedrola M O, Alattabi A W, Abdulredha M and Alawsh R 2019 Electrocoagulation as a green technology for phosphate removal from River water *Separation and Purification Technology* **210** 135-44
[27] Hashim K, Kot P, Zubaid S, Alwash R, Al Khaddar R, Shaw A, Al-Jumeily D and Aljefery M 2020 Energy efficient electrocoagulation using baffle-plates electrodes for efficient Escherichia Coli removal from Wastewater *Journal of Water Process Engineering* **33** 101079-86
[28] Hashim K S, Ewadh H M, Muhsin A A, Zubaidi S L, Kot P, Muradov M, Aljefery M and Al-Khaddar R 2020 Phosphate removal from water using bottom ash: Adsorption performance, coexisting anions and modelling studies *Water Science and Technology* **83** 1-17
[29] Abdulhadi B, Kot P, Hashim K, Shaw A, Muradov M and Al-Khaddar R 2021 Continuous-flow electrocoagulation (EC) process for iron removal from water: Experimental, statistical and economic study *Science of The Total Environment* **756** 1-16
[30] Hashim K S, Shaw A, AlKhaddar R, Kot P and Al-Shamma'a A 2021 Water purification from metal ions in the presence of organic matter using electromagnetic radiation-assisted treatment *Journal of Cleaner Production* **280**
[31] Hashim K S, Al-Saati N H, Alquzweeni S S, Zubaidi S L, Kot P, Kraidi L, Hussein A H, Alkhaddar R, Shaw A and Alwash R 2019 Decolourization of dye solutions by electrocoagulation: an investigation of the effect of operational parameters. In: *First International Conference on Civil and Environmental Engineering Technologies (ICCEET),* (University of Kufa, Iraq pp 25-32
[32] Alenezi A K, Hasan H A, Hashim K S, Amoako-Attah J, Gkantou M, Muradov M, Kot P and Abdulhadi B 2020 Zeolite-assisted electrocoagulation for remediation of phosphate from calcium-phosphate solution. In: *IOP Conference Series: Materials Science and Engineering*: IOP Publishing) p 012031
[33] Alhendal M, Nasir M J, Hashim K S, Amoako-Attah J, Al-Faluji D, Muradov M, Kot P and Abdulhadi B 2020 Cost-effective hybrid filter for remediation of water from fluoride. In: *IOP Conference Series: Materials Science and Engineering*: IOP Publishing) p 012038
[34] Al-Marri S, AlQuzweeni S S, Hashim K S, AlKhaddar R, Kot P, AlKizwini R S, Zubaidi S L and Al-Khafaji Z S 2020 Ultrasonic-Electrocoagulation method for nitrate removal from water. In: *IOP Conference Series: Materials Science and Engineering*: IOP Publishing) p 012073
[35] Hashim K S, Hussein A H, Zubaidi S L, Kot P, Kraidi L, Alkhaddar R, Shaw A and Alwash R 2019 Effect of initial pH value on the removal of reactive black dye from water by electrocoagulation (EC) method. In: *2nd International Scientific Conference,* (Al-Qadisiyah University, Iraq pp 12-22
[36] Alattabi A W, Harris C, Alkhaddar R, Alzeyadi A and Hashim K 2017 Treatment of Residential Complexes' Wastewater using Environmentally Friendly Technology *Procedia Engineering* **196** 792-9
[37] Alattabi A W, Harris C B, Alkhaddar R M, Hashim K S, Ortoneda-Pedrola M and Phipps D 2017 Improving sludge settleability by introducing an innovative, two-stage settling sequencing batch reactor *Journal of Water Process Engineering* **20** 207-16
[38] Alenazi M, Hashim K S, Hassan A A, Muradov M, Kot P and Abdulhadi B 2020 Turbidity removal using natural coagulants derived from the seeds of strychnos potatorum: statistical and experimental approach. In: *IOP Conference Series: Materials Science and Engineering*: IOP Publishing) p 012064

[39] Al-Jumeily D, Hashim K, Alkaddar R, Al-Tufaily M and Lunn J 2019 Sustainable and Environmental Friendly Ancient Reed Houses (Inspired by the Past to Motivate the Future). In: *11th International Conference on Developments in eSystems Engineering (DeSE),* (Cambridge, UK pp 214-9

[40] Omran I I, Al-Saati N H, Hashim K S, Al-Saati Z N, Patryk K, Khaddar R A, Al-Jumeily D, Shaw A, Ruddock F and Aljefery M 2019 Assessment of heavy metal pollution in the Great Al-Mussaib irrigation channel *Desalination and Water Treatment* **168** 165-74

[41] Abdulhadi B A, Kot P, Hashim K S, Shaw A and Khaddar R A 2019 Influence of current density and electrodes spacing on reactive red 120 dye removal from dyed water using electrocoagulation/electroflotation (EC/EF) process. In: *First International Conference on Civil and Environmental Engineering Technologies (ICCEET),* (University of Kufa, Iraq pp 12-22

[42] Idowu I A, Atherton W, Hashim K, Kot P, Alkhaddar R, Alo B I and Shaw A 2019 An analyses of the status of landfill classification systems in developing countries: Sub Saharan Africa landfill experiences *Waste Management* **87** 761-71

[43] Abdulredha M, Rafid A, Jordan D and Hashim K 2017 The development of a waste management system in Kerbala during major pilgrimage events: determination of solid waste composition *Procedia Engineering* **196** 779-84

[44] Abdulredha M, Al Khaddar R, Jordan D, Kot P, Abdulridha A and Hashim K 2018 Estimating solid waste generation by hospitality industry during major festivals: A quantification model based on multiple regression *Waste Management* **77** 388-400

[45] Hashim K S, Al-Saati N H, Hussein A H and Al-Saati Z N 2018 An investigation into the level of heavy metals leaching from canal-dreged sediment: a case study metals leaching from dreged sediment. In: *First International Conference on Materials Engineering & Science,* (Istanbul Aydın University (IAU), Turkey pp 12-22

[46] Hashim K S, Shaw A, Al Khaddar R, Ortoneda Pedrola M and Phipps D 2017 Defluoridation of drinking water using a new flow column-electrocoagulation reactor (FCER) - Experimental, statistical, and economic approach *Journal of Environmental Management* **197** 80-8

[47] Hashim K S, Shaw A, Al Khaddar R, Pedrola M O and Phipps D 2017 Energy efficient electrocoagulation using a new flow column reactor to remove nitrate from drinking water - Experimental, statistical, and economic approach *Journal of Environmental Management* **196** 224-33

[48] Hashim K S, Idowu I A, Jasim N, Al Khaddar R, Shaw A, Phipps D, Kot P, Pedrola M O, Alattabi A W and Abdulredha M 2018 Removal of phosphate from River water using a new baffle plates electrochemical reactor *MethodsX* **5** 1413-8

[49] Kadhim A, Sadique M, Al-Mufti R and Hashim K 2020 Long-term performance of novel high-calcium one-part alkali-activated cement developed from thermally activated lime kiln dust *Journal of Building Engineering* **32** 1-17

[50] Abdulredha M, Rafid A, Jordan D and Hashim K J P E 2017 The development of a waste management system in Kerbala during major pilgrimage events: determination of solid waste composition **196** 779-84

[51] Al-Anbari R, Alnakeeb A, Abdulredha M J E and Journal T 2013 Landfill site selection for Kerbala municipal solid wastes by using geographical information system techniques **32** 13

[52] Shubbar A A, Jafer H, Abdulredha M, Al-Khafaji Z S, Nasr M S, Al Masoodi Z and Sadique M J J o B E 2020 Properties of cement mortar incorporated high volume fraction of GGBFS and CKD from 1 day to 550 days **30** 101327

[53] Barnat-Hunek D, Góra J, Suchorab Z and Łagód G 2018 *Waste and Supplementary Cementitious Materials in Concrete,* pp 149-80

[54] Mehta A and Ashish D K 2020 Silica fume and waste glass in cement concrete production: A review *Journal of Building Engineering* **29**

[55] Gkantou M, Muradov M, Kamaris G S, Hashim K, Atherton W and Kot P 2019 Novel Electromagnetic Sensors Embedded in Reinforced Concrete Beams for Crack Detection *Sensors* **19** 5175-89

IOP Conf. Series: Materials Science and Engineering 1184 (2021) 012012 doi:10.1088/1757-899X/1184/1/012012

[56] Teng K H, Kot P, Muradov M, Shaw A, Hashim K, Gkantou M and Al-Shamma'a A 2019 Embedded Smart Antenna for Non-Destructive Testing and Evaluation (NDT&E) of Moisture Content and Deterioration in Concrete *Sensors* **19** 547-59

[57] Ryecroft S, Shaw A, Fergus P, Kot P, Hashim K, Moody A and Conway L 2019 A First Implementation of Underwater Communications in Raw Water Using the 433 MHz Frequency Combined with a Bowtie Antenna *Sensors* **19** 1813-23

[58] Ryecroft S P, shaw A, Fergus P, Kot P, Hashim K and Conway L 2019 A Novel Gesomin Detection Method Based on Microwave Spectroscopy. In: *12th International Conference on Developments in eSystems Engineering (DeSE),* (Kazan, Russia pp 429-33

The use of sawdust in the separation of copper from polluted water

M Al-Jannbi [1], M Almansorry [2,*]
[1] College of Engineering, University of Tikrit, Iraq.
[2] College of Engineering, University of Kerbala, Kerbala, Iraq.

Email: drmuna@yahoo.com

Abstract. Heavy metals in water supplies are a major source of concern around the world for a variety of reasons. For one, heavy metals can survive in water for lengthy stretches of time because they are not degraded by microorganisms. Second, heavy metals can build up to toxic levels in the humans and plants, posing a threat to living through the food. Presence of heavy metal in water causes serious health issues, including kidney failure and brain illnesses. As a result, the advancement of efficient water treatment technologies for heavy metal remediation is a top priority for scientists today. To extract copper, sawdust is used as a chemically active adsorbent in this study. The adsorption tests were done in a batch environment to see how various parameters like treating time, pH level, and adsorbent dosage affected the results. The results showed that the pH level that allowed the most copper to be removed was neutral. Following the batch trials, the optimal treating time was determined to be 165 minutes. The results showed that a 2.5 g/L dose of sawdust reduced copper pollution content in the water solution by 90 percent. Copper reduction has been successfully achieved using sawdust as a low-cost sorbent.

Keyword: Copper; water treatment; sawdust

1.Introduction

The pollution of heavy metal in water is regarded as the primary problem for wastewater treatment facilities around world because these metals can persist in water for lengthy time of time due to their inability to be degraded by microorganisms, and they can grow to harmful levels in the bodies, affecting humans by the food [1-4]. Furthermore, heavy metal contamination of water causes serious health issues for users, including kidney diseases, liver injury, brain disorders, and abortions[5-9]. Heavy metals like Cr, Cu, Se, As, Cd, and Ni, are toxic and cause diseases and disorders in living organisms [10-14].
Many factories, such as batteries, mines, tanneries, agrochemicals, and the paper industry, discharge heavy metals into the environment [3, 14-19]. These metals, which are considered the very dangerous, can affect humans even at low concentrations[20-23]. As a result, eliminating heavy metals from waterways is critical for human and wildlife protection. Copper is a radioactive heavy metal that accumulates in human bodies via the food chain and causes a variety of physiological problems [24-29]. Copper is used in a variety of industries, including fertilization, paint, dye processing, electroplating, and wood production [3, 14, 19, 30]. Copper concentrations of drinking water should not exceed 1.5 parts per million, according to the world health organization [30, 31]. Furthermore, China sets 0.5 ppm of copper concentration in industrial effluent,

according to GB25467-2010 [4, 32]. Increased copper concentrations may have severe toxic consequences when it affects the liver, skin, brain, and pancreas. This could result in kidney disease, headache, breathing problems, and nausea [19, 33, 34]. Electrocoagulation [35, 36], electrodialysis [37, 38], flotation [39], chemical precipitation [40], biological treatment [41-43] , adsorption [44-47], and coagulation [28] are some of the heavy metal removal strategies. Most of these methods necessitate a long treatment time and are either expensive or inefficient at high pollutant concentrations [2, 10]. The cost of commercial adsorption materials, for example, limits the adsorption technique. As a result, choosing sorbent is critical for developing a simple, environmentally sustainable, and low-cost system [3, 14, 19].
Researchers have extensively studied the adsorption method in their experiments to strip toxins from waterways by using various natural resources like animal waste and industrial by-products, according to previous literature [1, 14, 39]. Corn stalks, tree bark, tea leaves, and sawdust are among the most common materials used [22, 23]. The natural sorbent materials used in this study can be used with or without alteration to improve adsorption performance [22, 34, 48]. Sawdust, peats, straw, tree bark, and other woody materials are considered low-cost sorbents and the perfect substitute for industrially processed sorbents [4, 14]. The benefits of using these sorbents to remove heavy metals are determined by many factors, including their adsorption potential, recovery capability, and removal selectivity [33, 43, 49]. Sawdust is a waste product from the timber industry that can also be used as a cheap adsorbent for processing acids, unneeded compounds, and dyes. Sawdust's adsorption characteristic is influenced by its structure which contains phenolic acids, lignin, and cellulose[4, 29, 50-52]. Sawdust usage as cheap adsorbent to remove metals supports the wood business by getting rid of waste in addition to saving the forest. Because of the increased release of industrial and agricultural wastewaters containing high concentrations of heavy metals, as well as global warming, which affects water supply and contamination, human must find effective treatment methods to reduce the impact of the pollution[32, 33, 49, 53, 54].
Based on the above review of f research literature, the key goal of this research is to evaluate the efficiency of carbonized sawdust in terms of removing copper from polluted water bodies. It also looks at how operational conditions including treatment duration, pH, and sawdust dosages affect the results of copper removal from polluted water bodies.

2. Method

Sawdust was sprayed with water to clean and eliminate any dust or other chemicals. The washed sawdust dried by espousing it to direct sun light to implement it as activated carbon. Then, cleaned sawdust was chemically activated through combining 1 unit of weigh of sawdust with 2 units of weigh of concentrated sulfuric acid (H2SO4). At the end of this step the carbonized sawdust was generated. Yet, for drying purposes, the activated sawdust was dried in oven under to a temperature of 160°C for 24 hours. After that, the generated substance was cleaned several times with water before being dried at 105°C to make sure that the uncontacted H2SO4was removed.

The produced carbonized powder was used for copper adsorption in the present study. The adsorbent particles were between 0.6 and 0.10 mm in size. The specified volume of copper nitrates was dissolved in 300 mL of deionized water to generated water polluted with copper that contains 120 mg/L. The copper dilution from the stock solution was done properly. All the tests were carried out with a 150 mL copper solution and a specified dosage of sawdust. The polluted water were shook by a shaker at 200 revolution per minutes while the tests were carried out at a temperature of 20 ± 1 ◦C.

3. Discussion of the findings

3.1. *Influence of adsorbent dose*

Figure 1 depicts the effects of adsorbent dosages the removal performance of copper. The dosage ranged from 1 to 4 g/L. In the tests, the mixing speed was set to 200 revolution per minutes. Furthermore, a steady copper concentration of 5 mg/L was kept in the solutions with a neutral pH level and a contact time of 200 minutes. From the figure, the removal of the copper from the water increases with the increase in the dosage of the sawdust. The available surface for adsorption increases proportionally as the sawdust concentration rises in the water solution. As a result, the adsorption increased to about 100% for elevated adsorbent doses of four gram per liter. Therefore, the copper pollution in the water was reduced. Lower adsorbent concentrations result in lower adsorption and no further pollutant can be adsorbed. The best results were achieved with a dose of 2.5 g/L of sawdust and a copper removal percentage reached about 90%.

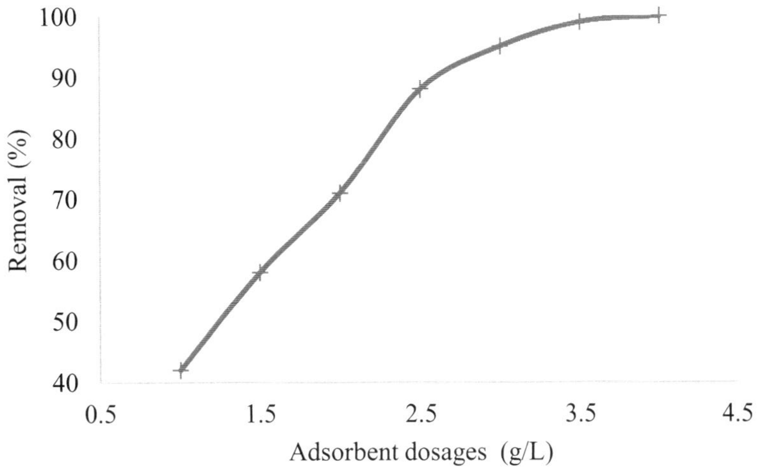

Figure 1: The relationship between copper removal and sawdust dosage.

3.2. Influence of treatment time

To comprehensively investigate copper elimination from polluted water, the effect of treatment time on the effectiveness of pollutant adsorption is investigated. The pH level of the water was 5, and the mixing speed was 200 revolutions per minute. The copper concentration was maintained at 5 ppm and the adsorbent dose of 2.5 ppm. Figure 2 depicts the relationship between touch time and copper removal from polluted water. The performance of the adsorbent improved with increasing treatment duration until saturation, as seen in Figure 1. As the duration extended from 20 to 165 minutes, the removal productivity increased from 30% to 70 percent. The removal efficiency was about 70 percent between both 165 and 200 minutes of treatment time. As a result, it was recommended that 165 minutes of treatment time be included in further investigation in this study.

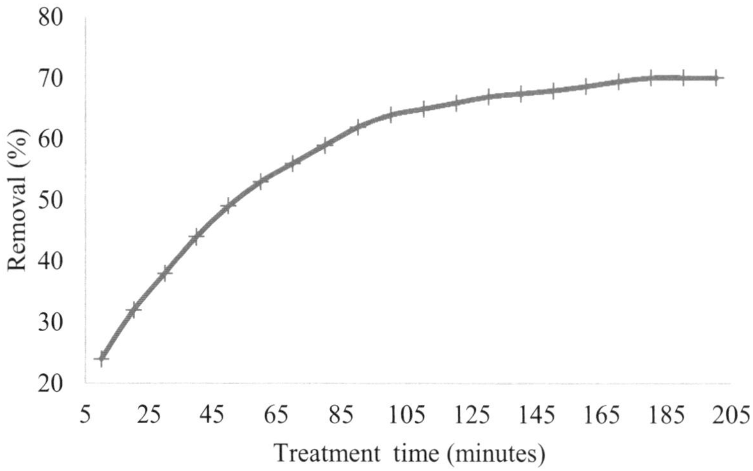

Figure 2: The relationship between copper removal and treatment time.

3.3. Influence of solution pH level

The level of the pH is thought to be a significant factor affecting metal adsorption, and is confirmed by previous research [2, 10]. This is because the useful clusters of the media and the chemical composition of the metal are affected by the pH of an aqueous solution. A pH range of 2 to 9 has been used in the experiments of this research the experiments to explore the influence of pH on adsorption preference. The sulfuric acid (H2SO40) was implemented to change the pH level of the polluted solution. The concentration of the copper maintained at 5 ppm, and the treatment duration was set to 165 minutes. Furthermore, the adsorbent dose was 2.5 ppm, with mixing speed at 200 revolutions per minute. Copper removal efficiencies vs. pH values are depicted in Figure 3. Adsorption improved as the pH value was raised from 2 to 6. After 7, the removal efficiencies declined marginally. In pH 7, the maximal removal was 81 percent. As the adsorbent surface allowed positively charged ion to be attached, removal performance was poor at acidic levels. That can be explained by the increased rivalry between H+ and the ions of the copper for usable adsorption sites. Removal efficiency began to decline slightly at pH 8 and 9, which may be due to the development of soluble hydroxyl complexes and $Cu(OH)_2$.

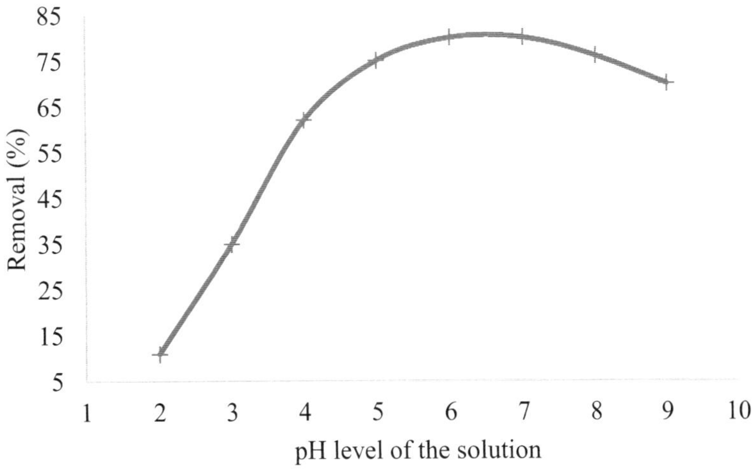

Figure 3: The relationship between copper removal and pH level.

Several recent reports questioned the adsorption methods' cost-effectiveness and accuracy because used adsorption media have to be disposed of in disposal areas, which have elevated building and repair expenses [4, 9, 29, 49-53, 55-59]. These disadvantages may be overcome by recycling used adsorbent in building practices, like those described in the previous researches [60-65], and improving precision by using computerized sensors [66-69].

4. Conclusion

The sawdust which is considered as cheap material, has been tested for in this research as adsorption and non-extensive copper removal process. Copper adsorption is heavily dependent on the length of contact between adsorbent and pollutants, the amount of sawdust used, and the pH level. As the pH of the solution is increased from 1 to 6, the copper elimination from wastewater increases. However, after 7, the removal efficiency decreased. The best pH level for copper removal was at neutral pH, and best treatment time for copper removal was deemed to be 165 minutes. It has been revealed that raising the dose of adsorbent improved the treatment degree due to the accessible are for adsorption increased. A 2.5 g/L dosage of sawdust reduced the copper content in water by 90 percent. This research showed that the Sawdust is very good sorbent for removing copper from water. Research is required to solve the key disadvantages of adsorption processes, where several previous studies questioned their cost-effectiveness and accuracy.

References

[1] Hashim K S, Al-Saati N H, Hussein A H and Al-Saati Z N 2018 An investigation into the level of heavy metals leaching from canal-dreged sediment: a case study metals leaching from dreged sediment. In: *First International Conference on Materials Engineering & Science,* (Istanbul Aydın University (IAU), Turkey pp 12-22

[2] Abdulraheem F S, Al-Khafaji Z S, Hashim K S, Muradov M, Kot P and Shubbar A A 2020 Natural filtration unit for removal of heavy metals from water. In: *IOP Conference Series: Materials Science and Engineering*: IOP Publishing) p 012034

[3] Jawad S F, Saddam N S, Adaami Q J, Kareem M M, Abdulredha M, Mubarak H A, Kot P, Gkantou M and AlKhayyat A 2021 Dye removal from textile wastewater using solar-powered

electrocoagulation reactor. In: *IOP Conference Series: Materials Science and Engineering*: IOP Publishing) p 012016

[4] Abdulredha M, Kadhim N, Hussein A, Almutairi M, Alkhaddar R, Yeboah D and Hashim K 2021 Zeolite as a natural adsorbent for nitrogenous compounds being removed from water. In: *IOP Conference Series: Materials Science and Engineering*: IOP Publishing) p 012082

[5] Aayef A N, Al Masoodi W T M, Kamel R J, Abdulredha M, Almansoori N A, Kot P and Muradov M 2021 An experimental study for adapting electrocoagulation as a technique for fluoride removal from water. In: *IOP Conference Series: Materials Science and Engineering*: IOP Publishing) p 012012

[6] Abdulredha M, Abdulridha A, Shubbar A, Alkhaddar R, Kot P and Jordan D 2020 Estimating municipal solid waste generation from service processions during the Ashura religious event. In: *IOP Conference Series: Materials Science and Engineering*: IOP Publishing) p 012075

[7] Abdulredha M, Al Khaddar R and Jordan D 2017 Hoteliers' attitude towards solid waste source separation through mega festivals: A pilot study in Karbala. In: *International Conference for Doctoral Research*: BUID)

[8] Abdulredha M, Al Khaddar R, Jordan D, Al-Attabi A and Alzeyadi A 2017 Public participation in solid waste management during mega festivals: A pilot study. In: *WCST World Congress on Sustainable Technologies Proceedings 2017*: Infonomics Society) pp 38-41

[9] Abdulredha M, Al Khaddar R, Jordan D, Kot P, Abdulridha A and Hashim K J W M 2018 Estimating solid waste generation by hospitality industry during major festivals: A quantification model based on multiple regression **77** 388-400

[10] Abdulla G, Kareem M M, Hashim K S, Muradov M, Kot P, Mubarak H A, Abdellatif M and Abdulhadi B 2020 Removal of iron from wastewater using a hybrid filter. In: *IOP Conference Series: Materials Science and Engineering*: IOP Publishing) p 012035

[11] Hashim K S, Shaw A, Al Khaddar R, Pedrola M O and Phipps D 2017 Iron removal, energy consumption and operating cost of electrocoagulation of drinking water using a new flow column reactor *Journal of Environmental Management* **189** 98-108

[12] Al-Saati N H, Hussein T K, Abbas M H, Hashim K, Al-Saati Z N, Kot P, Sadique M, Aljefery M H and Carnacina I 2019 Statistical modelling of turbidity removal applied to non-toxic natural coagulants in water treatment: a case study *Desalination and Water Treatment* **150** 406-12

[13] Omran I I, Al-Saati N H, Hashim K S, Al-Saati Z N, Patryk K, Khaddar R A, Al-Jumeily D, Shaw A, Ruddock F and Aljefery M 2019 Assessment of heavy metal pollution in the Great Al-Mussaib irrigation channel *Desalination and Water Treatment* **168** 165-74

[14] Isra'a S S, Al-Janabi A, Abdulredha M, Alkandari A, Abdellatif M and Yeboah D 2021 Reusing of furnace bottom ash as an adsorbent for phosphate removal from water. In: *IOP Conference Series: Materials Science and Engineering*: IOP Publishing) p 012006

[15] Hashim K S, Idowu I A, Jasim N, Al Khaddar R, Shaw A, Phipps D, Kot P, Pedrola M O, Alattabi A W and Abdulredha M 2018 Removal of phosphate from River water using a new baffle plates electrochemical reactor *MethodsX* **5** 1413-8

[16] Abdulhadi B A, Kot P, Hashim K S, Shaw A and Khaddar R A 2019 Influence of current density and electrodes spacing on reactive red 120 dye removal from dyed water using electrocoagulation/electroflotation (EC/EF) process. In: *First International Conference on Civil and Environmental Engineering Technologies (ICCEET),* (University of Kufa, Iraq pp 12-22

[17] Hashim K S, Hussein A H, Zubaidi S L, Kot P, Kraidi L, Alkhaddar R, Shaw A and Alwash R 2019 Effect of initial pH value on the removal of reactive black dye from water by electrocoagulation (EC) method. In: *2nd International Scientific Conference,* (Al-Qadisiyah University, Iraq pp 12-22

[18] Grmasha R A, Al-sareji O J, Salman J M, Hashim K S and Jasim I A 2020 Polycyclic Aromatic Hydrocarbons (PAHs) in Urban Street Dust WithinThree Land-Uses of Babylon Governorate, Iraq:

Distribution, Sources, andHealth Risk Assessment *Journal of King Saud University - Engineering Sciences* **33** 1-18

[19] Hashim K S, Idowu I A, Jasim N, Al Khaddar R, Shaw A, Phipps D, Kot P, Pedrola M O, Alattabi A W and Abdulredha M J M 2018 Removal of phosphate from River water using a new baffle plates electrochemical reactor **5** 1413-8

[20] Abdulhadi B, Kot P, Hashim K, Shaw A, Muradov M and Al-Khaddar R 2021 Continuous-flow electrocoagulation (EC) process for iron removal from water: Experimental, statistical and economic study *Science of The Total Environment* **756** 1-16

[21] Hashim K S, Khaddar R A, Jasim N, Shaw A, Phipps D, Kot P, Pedrola M O, Alattabi A W, Abdulredha M and Alawsh R 2019 Electrocoagulation as a green technology for phosphate removal from River water *Separation and Purification Technology* **210** 135-44

[22] Alenazi M, Hashim K S, Hassan A A, Muradov M, Kot P and Abdulhadi B 2020 Turbidity removal using natural coagulants derived from the seeds of strychnos potatorum: statistical and experimental approach. In: *IOP Conference Series: Materials Science and Engineering*: IOP Publishing) p 012064

[23] Alenezi A K, Hasan H A, Hashim K S, Amoako-Attah J, Gkantou M, Muradov M, Kot P and Abdulhadi B 2020 Zeolite-assisted electrocoagulation for remediation of phosphate from calcium-phosphate solution. In: *IOP Conference Series: Materials Science and Engineering*: IOP Publishing) p 012031

[24] Alhendal M, Nasir M J, Hashim K S, Amoako-Attah J, Al-Faluji D, Muradov M, Kot P and Abdulhadi B 2020 Cost-effective hybrid filter for remediation of water from fluoride. In: *IOP Conference Series: Materials Science and Engineering*: IOP Publishing) p 012038

[25] Al-Marri S, AlQuzweeni S S, Hashim K S, AlKhaddar R, Kot P, AlKizwini R S, Zubaidi S L and Al-Khafaji Z S 2020 Ultrasonic-Electrocoagulation method for nitrate removal from water. In: *IOP Conference Series: Materials Science and Engineering*: IOP Publishing) p 012073

[26] Alyafei A, AlKizwini R S, Hashim K S, Yeboah D, Gkantou M, Al Khaddar R, Al-Faluji D and Zubaidi S L 2020 Treatment of effluents of construction industry using a combined filtration-electrocoagulation method. In: *IOP Conference Series: Materials Science and Engineering*: IOP Publishing) p 012032

[27] Hashim K S, Shaw A, Al Khaddar R, Ortoneda Pedrola M and Phipps D 2017 Defluoridation of drinking water using a new flow column-electrocoagulation reactor (FCER) - Experimental, statistical, and economic approach *Journal of Environmental Management* **197** 80-8

[28] Hashim K S, Shaw A, Al Khaddar R, Pedrola M O and Phipps D 2017 Energy efficient electrocoagulation using a new flow column reactor to remove nitrate from drinking water - Experimental, statistical, and economic approach *Journal of Environmental Management* **196** 224-33

[29] Abdulredha M, Rafid A, Jordan D and Alattabi A J P e 2017 Facing up to waste: how can hotel managers in Kerbala, Iraq, help the city deal with its waste problem? **196** 771-8

[30] Jaishankar M, Tseten T, Anbalagan N, Mathew B B and Beeregowda K N 2014 Toxicity, mechanism and health effects of some heavy metals *Interdisciplinary toxicology* **7** 60-72

[31] Aqeel K, Mubarak H A, Amoako-Attah J, Abdul-Rahaim L A, Al Khaddar R, Abdellatif M, Al-Janabi A and Hashim K S 2020 Electrochemical removal of brilliant green dye from wastewater. In: *IOP Conference Series: Materials Science and Engineering*: IOP Publishing) p 012036

[32] Al-Sareji O J, Abdulredha M, Mubarak H A, Grmasha R A, Alnowaishry A, Kot P, Al-Khaddar R and AlKhayyat A 2021 Copper removal from water using carbonized sawdust. In: *IOP Conference Series: Materials Science and Engineering*: IOP Publishing) p 012015

[33] Alattabi A W, Harris C, Alkhaddar R, Alzeyadi A and Abdulredha M J P e 2017 Online monitoring of a sequencing batch reactor treating domestic wastewater **196** 800-7

[34] Shubbar A A, Jafer H, Abdulredha M, Al-Khafaji Z S, Nasr M S, Al Masoodi Z and Sadique M J J o B E 2020 Properties of cement mortar incorporated high volume fraction of GGBFS and CKD from 1 day to 550 days **30** 101327

[35] Emamjomeh M M, Mousazadeh M, Mokhtari N, Jamali H A, Makkiabadi M, Naghdali Z, Hashim K S and Ghanbari R 2020 Simultaneous removal of phenol and linear alkylbenzene sulfonate from automotive service station wastewater: Optimization of coupled electrochemical and physical processes *Separation Science and Technology* **55** 3184-94

[36] Hashim K S, AlKhaddar R, Shaw A, Kot P, Al-Jumeily D, Alwash R and Aljefery M H 2020 *Electrocoagulation as an eco-friendly River water treatment method. In Advances in Water Resources Engineering and Management* (Berline: Springer)

[37] Hashim K S, Al-Saati N H, Alquzweeni S S, Zubaidi S L, Kot P, Kraidi L, Hussein A H, Alkhaddar R, Shaw A and Alwash R 2019 Decolourization of dye solutions by electrocoagulation: an investigation of the effect of operational parameters. In: *First International Conference on Civil and Environmental Engineering Technologies (ICCEET),* (University of Kufa, Iraq pp 25-32

[38] Mohammed A-H, Hussein A H, Yeboah D, Al Khaddar R, Abdulhadi B, Shubbar A A and Hashim K S 2020 Electrochemical removal of nitrate from wastewater. In: *IOP Conference Series: Materials Science and Engineering*: IOP Publishing) p 012037

[39] Hashim K, Kot P, Zubaid S, Alwash R, Al Khaddar R, Shaw A, Al-Jumeily D and Aljefery M 2020 Energy efficient electrocoagulation using baffle-plates electrodes for efficient Escherichia Coli removal from Wastewater *Journal of Water Process Engineering* **33** 101079-86

[40] Zanki A K, Mohammad F H, Hashim K S, Muradov M, Kot P, Kareem M M and Abdulhadi B 2020 Removal of organic matter from water using ultrasonic-assisted electrocoagulation method. In: *IOP Conference Series: Materials Science and Engineering*: IOP Publishing) p 012033

[41] Alattabi A W, Harris C, Alkhaddar R, Alzeyadi A and Hashim K 2017 Treatment of Residential Complexes' Wastewater using Environmentally Friendly Technology *Procedia Engineering* **196** 792-9

[42] Alattabi A W, Harris C B, Alkhaddar R M, Hashim K S, Ortoneda-Pedrola M and Phipps D 2017 Improving sludge settleability by introducing an innovative, two-stage settling sequencing batch reactor *Journal of Water Process Engineering* **20** 207-16

[43] Hashim K S, Al Khaddar R, Jasim N, Shaw A, Phipps D, Kot P, Pedrola M O, Alattabi A W, Abdulredha M, Alawsh R J S and Technology P 2019 Electrocoagulation as a green technology for phosphate removal from River water **210** 135-44

[44] Hashim K S, Ali S S M, AlRifaie J K, Kot P, Shaw A, Al Khaddar R, Idowu I and Gkantou M 2020 Escherichia coli inactivation using a hybrid ultrasonic–electrocoagulation reactor *Chemosphere* **247** 125868-75

[45] Hassan Alnaimi I J I, Abuduljaleel Al-Janabi, Khalid Hashim, Michaela Gkantou, Salah L. Zubaidi, Patryk Kot, Magomed Muradov 2020 Ultrasonic-electrochemical treatment for effluents of concrete plants Ultrasonic-electrochemical treatment for effluents of concrete plants. In: *IOP Conference Series Materials Science and Engineering* (University of Kufa, Najaf, Iraq pp 1-9

[46] Hashim K S, Shaw A, AlKhaddar R, Kot P and Al-Shamma'a A 2021 Water purification from metal ions in the presence of organic matter using electromagnetic radiation-assisted treatment *Journal of Cleaner Production* **280**

[47] Hashim K S, Ewadh H M, Muhsin A A, Zubaidi S L, Kot P, Muradov M, Aljefery M and Al-Khaddar R 2020 Phosphate removal from water using bottom ash: Adsorption performance, coexisting anions and modelling studies *Water Science and Technology* **83** 1-17

[48] Majdi H S, Shubbar A, Nasr M S, Al-Khafaji Z S, Jafer H, Abdulredha M, Al Masoodi Z, Sadique M and Hashim K J D i B 2020 Experimental data on compressive strength and ultrasonic pulse velocity properties of sustainable mortar made with high content of GGBFS and CKD combinations

[49] Abdulredha M, Rafid A, Jordan D and Hashim K J P E 2017 The development of a waste management system in Kerbala during major pilgrimage events: determination of solid waste composition **196** 779-84

[50] Abdulredha M, al-Khaddar R, Kot P, Jordan D and Abdulridha A 2018 Benchmarking of the Current Solid Waste Management System in Karbala, Iraq, Using Wasteaware Benchmark Indicators. In: *World Environmental and Water Resources Congress 2018: Groundwater, Sustainability, and Hydro-Climate/Climate Change*: American Society of Civil Engineers Reston, VA) pp 40-8

[51] Abdulredha M, Kot P, Al Khaddar R, Jordan D, Abdulridha A J E, Development and Sustainability 2020 Investigating municipal solid waste management system performance during the Arba'een event in the city of Kerbala, Iraq **22** 1431-54

[52] Abdulredha M, Muhsin A A, Al-Janabi A, Alajmi B N, Gkantou M, Amoako-Attah J, Al-Jumeily D, Mustafina J and AlKhayyat A 2021 Using SF and CKD as cement replacement materials for producing cement mortar. In: *IOP Conference Series: Materials Science and Engineering*: IOP Publishing) p 012007

[53] Al-Anbari R, Alnakeeb A, Abdulredha M J E and Journal T 2013 Landfill site selection for Kerbala municipal solid wastes by using geographical information system techniques **32** 13

[54] ALWAN H H, SALEH L A, AL-MOHAMMED F M, ABDULREDHA M A J J o E S and Technology 2020 EXPERIMENTAL PREDICTION OF THE DISCHARGE COEFFICIENTS FOR RECTANGULAR WEIR WITH BOTTOM ORIFICES **15** 3265-80

[55] Idowu I A, Atherton W, Hashim K, Kot P, Alkhaddar R, Alo B I and Shaw A 2019 An analyses of the status of landfill classification systems in developing countries: Sub Saharan Africa landfill experiences *Waste Management* **87** 761-71

[56] Abdulredha M, Rafid A, Jordan D and Alattabi A 2017 Facing up to waste: how can hotel managers in Kerbala, Iraq, help the city deal with its waste problem? *Procedia engineering* **196** 771-8

[57] Abdulredha M, Kot P, Al Khaddar R, Jordan D and Abdulridha A 2020 Investigating municipal solid waste management system performance during the Arba'een event in the city of Kerbala, Iraq *Environment, Development and Sustainability* **22** 1431-54

[58] Abdulredha M, Rafid A, Jordan D and Hashim K 2017 The development of a waste management system in Kerbala during major pilgrimage events: determination of solid waste composition *Procedia Engineering* **196** 779-84

[59] Abdulredha M, Al Khaddar R, Jordan D, Kot P, Abdulridha A and Hashim K 2018 Estimating solid waste generation by hospitality industry during major festivals: A quantification model based on multiple regression *Waste Management* **77** 388-400

[60] Shubbar A A, Jafer H, Dulaimi A, Hashim K, Atherton W and Sadique M 2018 The development of a low carbon binder produced from the ternary blending of cement, ground granulated blast furnace slag and high calcium fly ash: An experimental and statistical approach *Construction and Building Materials* **187** 1051-60

[61] Shubbar A A, Al-Shaer A, AlKizwini R S, Hashim K, Hawesah H A and Sadique M 2019 Investigating the influence of cement replacement by high volume of GGBS and PFA on the mechanical performance of cement mortar. In: *First International Conference on Civil and Environmental Engineering Technologies (ICCEET),* (University of Kufa, Iraq pp 31-8

[62] Kadhim A, Sadique M, Al-Mufti R and Hashim K 2020 Long-term performance of novel high-calcium one-part alkali-activated cement developed from thermally activated lime kiln dust *Journal of Building Engineering* **32** 1-17

[63] Kadhim A, Sadique M, Al-Mufti R and Hashim K 2020 Developing One-Part Alkali-Activated metakaolin/natural pozzolan Binders using Lime Waste as activation Agent *Advances in Cement Research* **32** 1-38

[64] Shubbar A A, Sadique M, Nasr M S, Al-Khafaji Z S and Hashim K S 2020 The impact of grinding time on properties of cement mortar incorporated high volume waste paper sludge ash *Karbala International Journal of Modern Science* **6** 1-23

[65] Shubbar A A, Sadique M, Shanbara H K and Hashim K 2020 *The Development of a New Low Carbon Binder for Construction as an Alternative to Cement. In Advances in Sustainable Construction Materials and Geotechnical Engineering* (Berlin: Springer)

[66] Gkantou M, Muradov M, Kamaris G S, Hashim K, Atherton W and Kot P 2019 Novel Electromagnetic Sensors Embedded in Reinforced Concrete Beams for Crack Detection *Sensors* **19** 5175-89

[67] Ryecroft S, Shaw A, Fergus P, Kot P, Hashim K, Moody A and Conway L 2019 A First Implementation of Underwater Communications in Raw Water Using the 433 MHz Frequency Combined with a Bowtie Antenna *Sensors* **19** 1813-23

[68] Ryecroft S P, shaw A, Fergus P, Kot P, Hashim K and Conway L 2019 A Novel Gesomin Detection Method Based on Microwave Spectroscopy. In: *12th International Conference on Developments in eSystems Engineering (DeSE),* (Kazan, Russia pp 429-33

[69] Teng K H, Kot P, Muradov M, Shaw A, Hashim K, Gkantou M and Al-Shamma'a A 2019 Embedded Smart Antenna for Non-Destructive Testing and Evaluation (NDT&E) of Moisture Content and Deterioration in Concrete *Sensors* **19** 547-59

Air Quality Analyses in the City of Karbala, Iraq

Rayed Abdullatif[1], Salah R Hammadi[1], Qassim Alsaady[1,*]
[1] Environment Department, Ministry of Oil, Iraq

Email: dr_qassim@gmail.com

Abstract. The pollution of the air is the most dangerous type of environmental contamination as the air volume on this planet is very low and people are not able to live in the absence of air for more than a few minutes. This research concentrates on analyzing the concatenation of two pollutants (PM_{10} and $PM_{2.5}$) in the city of Kerbala, Iraq. This has been achieved by building three surveillance stations on one main street in Kerbala. This street is considered is very vital in the city that links many districts and contains several governmental and health institutions such as hospitals. The Temp-Monitor model Airing-1000 is implemented the achieve the goal of this research. It has been recognized that the spectrum levels of the PM_{10} and the $PM_{2.5}$ were between 165.5 and 23.5 μg/m^3 and between 32 and 8.1 μg/m^3, respectively over the time of the investigation which begins by February 2020 and ends by July 2020. The average PM_{10} level from station one exceeds the annual level set by the WHO and the average $PM_{2.5}$ level is below the WHO annual standard, but the $PM_{2.5}$ level was greater than the WHO limit from station 2 in the middle of April 2020. Many elements help increase PM levels in urban areas, like emissions from cars, season, and other events. In this regard, it has been distinguished that all components play a critical part in the development of PM concentrations.

Keyword: Air contamination; PM_{10}; $PM_{2.5}$

1. Introduction

The population of the earth is exposed to many sources of contamination, including soil, air, and water pollution. All types of pollutants are recognized to directly impact public health and the Earth's ecosystem [1-4]. Yet, air pollution is the most dangerous type of contamination since the quantity of air on this planet is extremely limited, and people are not able to live without clean air for more than a few minutes [5, 6]. The application of effective water treatment techniques[7-9] like physical and chemical processing [10-16] or the application of restricted wastewater limitations [17-19], may lead to a significant degree of monitoring of water contamination. Moreover, living marine animals and vegetations contribute to the equilibrium of water pollutants [20-25]. Likewise, the use of effective treatment methods [26-30] will regulate soil contamination resulting from waste landfills [31-34] or the injecting of toxins into the soil [35-39].

On the other hand, air contamination is classified as the most harmful source of pollution on the planet owing to its small amount, and all beings are not able to live with little air over a few minutes [5, 40-42].

Even air emissions could come from natural causes like volcanoes, the main sources of air pollution come from human activities. For example, a large volume of particulates and other gases [43-45] are released into the air from the cement industry only in addition to huge quantities of carbon dioxide gas [46-48]. Moreover, human-made causes produce large amounts of greenhouse gases, which cause global warming [49-53] that causes extreme eco-pollution and water scarcity in many nations around the globe [54-58]. Overall, air pollution can be caused by dangerous suspended substances. Atmospheric contaminants like aerodynamically diametric particulate matter, particularly PM_{10} and $PM_{2,5}$ (particulates that have diameters smaller than less than 10 and 2,5 μm), can impact humans health causing illness and death even with exposure to very low levels of such pollutants.
Extreme negative effects are caused by, where about 2.1 additional deaths caused by particulates [59, 60]. Particulates with a 10 μm dimeter are capable of depositing in the higher side of the air tract in the lung [59, 61]. On the other hand, particulates with 2.5 μm diameter (small particulate matter) can stay in the lungs for an extended period with the potential to infiltrate the bloodstream. Several health effects like cancer, airborne disease, abdominal cancer, or coronary thrombosis diseases have recognized in terms of toxicological research. Particulates are a blend of chemicals, solids and liquids that have various sizes and diameters. are restricted chemical mixtures in the form of various solids and liquids. Most particulates contain are inorganic chemicals like chloride and nitrates, organic chemicals, and such as aromatic hydrocarbons.
There is controversy in the relationship between the type of land and particulates concentration. The research found that the relationship between particulates and residential areas was positive [59, 62-64]. On the other hand, other researchers showed that the levels of particulates in the same areas it low. However, many showed the connection between concentrations of particulates and the size of the industrial areas is positive. Several studies have shown that high levels of particulates in industrial and other areas of green fields have found low particulates levels. It has been reported that heavy traffic could lead to higher levels of particulates in many areas. However, very little researches linked city development with the concentration of particulates in the environment and the relationship between these two aspects were found to be diverse. To give an example, Weichenthal, Farrell, Goldberg, Joseph and Hatzopoulou [65] reported various connection between particulates and development.
Many researchers have studied the concentration of particulates matters in connection with the built environment around the world. This research aims to examine the concertation of two types of particulates matters that are PM_{10} and $PM_{2.5}$ in one of the main streets in the city of Karbala, Iraq that is Maytham Al-Tammar road. This road is considered the busiest rod in the city which connects the northern districts with the southern district.

2. Research Methodology

2.1. Kerbala City
The city of Karbala is positioned in the centre of Iraq and about 90 kilometres miles from Baghdad. It has an area of about 5 thousand square kilometres which represent about 1.2% of Iraq total area. Kerbala is inhabited by more than 1.2 million residents and host millions during religious events like the Ashura and the Arba'een. The Maytham Al-Tammar road was selected in this research to examine the concentration of the particulates namely PM_{10} and $PM_{2.5}$ for several reasons. First, the street is considered as one of the main roads in the city that links the northern districts with the southern districts of the city. Besides, the road is the busiest in the city as the majority of the city visitors goes to that road to be closest to the holy shrines in the city. Additionally, The Maytham Al-Tammar road contains more than 20 car garages and many commercial and institutional buildings like hospitals.

2.2. Collection the Data

IOP Conf. Series: Materials Science and Engineering 1184 (2021) 012014 doi:10.1088/1757-899X/1184/1/012014

The length of the Maytham Al-Tammar road is about 200 meters. Two monitoring stations were positioned in the road to monitor the concentration of the particles matter that are MP_{10} and $MP_{2.5}$ in the road. The first monitoring station is located near Karbala-Baghdad road while the second monitoring station is located in the middle of the Hamza the al-Zaghair street. Figure 1 shows the locations of the two monitoring stations on the road. Temtop monitor model Airing-1000 has been used in this research to monitor the concentration of particulates matter in the road. The accuracy of the Temtop monitor is about 0.1 μg/m and the detection range is 0 to 1000 μg/m. The manual of the detector was followed for operation correction operation and monitoring. This includes maintaining the temperature between 0 and 50 ^{0}C and humidity between 10% and 90%. The data collection started in February 2020 until July 2020 with an average of 10 readings per month in both stations. The selection of the duration is mainly based on the representation of the dry and wet seasons.

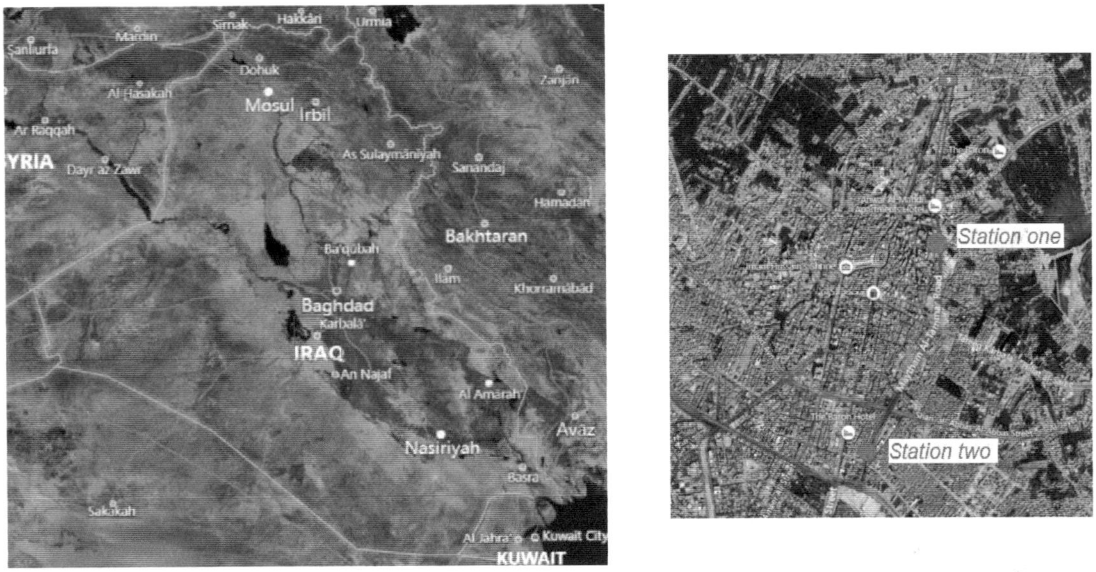

Figure 1: The locations of the monitoring station in Maytham Al-Tammar road.

3. Outcomes

The concentration readings of the MP_{10} are presented in Figure 1 for the Maytham Al-Tammar road in Kerbala city for the duration between February and July of 2020. Al the concentration readings are presented in μg/m^3. The concentration of the particulates matters during the study period ranged from 60 to 34 μg/m^3 with an average of 18.6 μg/m^3 for station one and 21.86 μg/m^3 for station two. The highest concentration of the MP_{10} was observed during July of 31 μg/m^3 and 34 μg/m^3 for one and two, respectively. The level of the MP_{10} is significantly higher than the standard set by the world health organization of 20 μg/m^3. This could cause serious health effects on the public in the area. However, the concentration of the MP_{10} was largely lower than the limit during February and March in the same year.

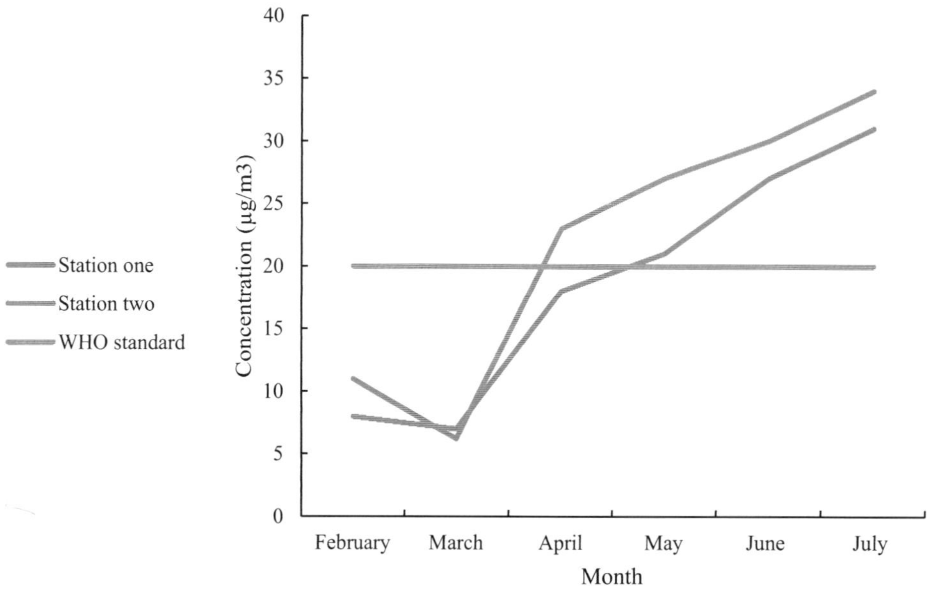

Figure 2: The concentration of the MP_{10} in the Maytham Al-Tammar road, Karbala, Iraq.

On the other hand, Figure 3 presents the concentration readings of the $MP_{2.5}$ in the Maytham Al-Tammar road in Kerbala city for the duration between February and July of 2020. Figure 3 clearly shows that the range of the $MP_{2.5}$ is between 4 and 10.8 $\mu g/m^3$ throughout the study. Like MP_{10}, the concentration of the $MP_{2.5}$ is lowest during February and March and the concentration of particulates increases during the summer months. The average concentration of the $MP_{2.5}$ was observed during May which was about 9.6 $\mu g/m^3$. The concentration of the $MP_{2.5}$ in station one and station two were comparable over the month of the study. The concertation increased of both particulate matters in the populated places is reported to be attributed to pollution from vehicles, which have also been supported by others like Weichenthal, Farrell, Goldberg, Joseph and Hatzopoulou [65]. Besides pollution from automobiles, it was noted that the recorded pollution in the city of Karbala was lower in the beginning study duration than at the end of the study duration because the precipitation during February and March is significantly higher than in June and July. The higher precipitation has led to a significant decrease in the concentration of the particulates.

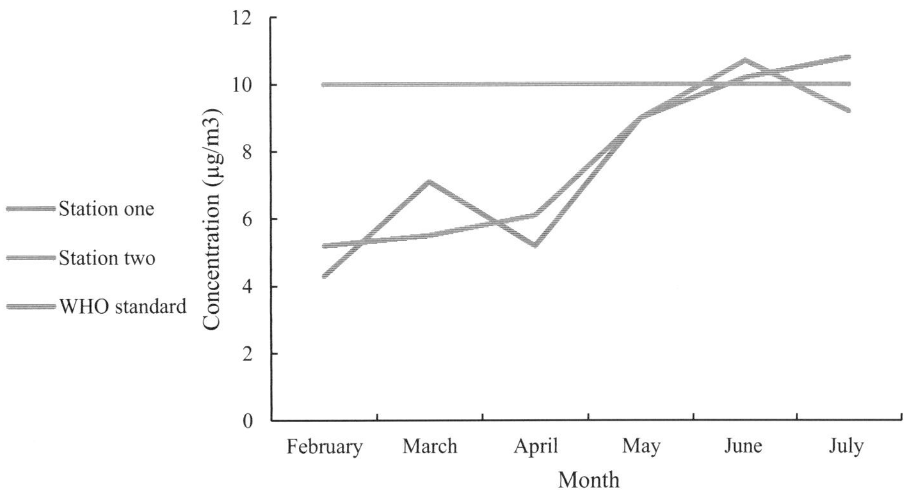

Figure 3: The concentration of the $MP_{2.5}$ in the Maytham Al-Tammar road, Karbala, Iraq.

4. Conclusion

The goal of this research is to assess air pollution in one of the main streets in the city of Karbala, Iraq. the Maytham Al-Tammar road in the city for the duration between February and July of 2020. The concentration of two air pollutants that are MP_{10} and $MP_{2.5}$ has been examined in this research. The selected road is considered one of the main roads in the city. The outcome of the research showed that the concentration of the MP_{10} is higher than the concentration level set by the WHO while the concentration level of the $MP_{2.5}$ is lower than the WHO standard. The range of the $MP_{2.5}$ is between 4 and 10.8 $\mu g/m^3$ over the duration of the study. on the other hand, the concentration of the particulates matters during the study period ranged from 60 to 34 $\mu g/m^3$ with an average of 18.6 $\mu g/m^3$ for station one and 21.86 $\mu g/m^3$ for station two. The concertation increased of both particulate matters in the populated places is reported to be attributed to pollution from vehicles.

References

[1] Al-Sareji O J, Abdulredha M, Mubarak H A, Grmasha R A, Alnowaishry A, Kot P, Al-Khaddar R and AlKhayyat A 2021 Copper removal from water using carbonized sawdust. In: *IOP Conference Series: Materials Science and Engineering*: IOP Publishing) p 012015

[2] ALWAN H H, SALEH L A, AL-MOHAMMED F M, ABDULREDHA M A J J o E S and Technology 2020 EXPERIMENTAL PREDICTION OF THE DISCHARGE COEFFICIENTS FOR RECTANGULAR WEIR WITH BOTTOM ORIFICES **15** 3265-80

[3] Farhan S L, Jasim I A and Al-Mamoori S K 2019 The transformation of the city of Najaf, Iraq: Analysis, reality and future prospects *Journal of Urban Regeneration & Renewal* **13** 160-71

[4] Hashim K S, Al Khaddar R, Jasim N, Shaw A, Phipps D, Kot P, Pedrola M O, Alattabi A W, Abdulredha M, Alawsh R J S and Technology P 2019 Electrocoagulation as a green technology for phosphate removal from River water **210** 135-44

[5] Zhu F, Ding R, Lei R, Cheng H, Liu J, Shen C, Zhang C, Xu Y, Xiao C and Li X 2019 The short-term effects of air pollution on respiratory diseases and lung cancer mortality in Hefei: A time-series analysis *Respiratory Medicine* **146** 57-65

[6] Grmasha R A, Al-sareji O J, Salman J M, Hashim K S and Jasim I A 2020 Polycyclic Aromatic Hydrocarbons (PAHs) in Urban Street Dust WithinThree Land-Uses of Babylon Governorate, Iraq: Distribution, Sources, andHealth Risk Assessment *Journal of King Saud University - Engineering Sciences* **33** 1-18

[7] Omran I I, Al-Saati N H, Hashim K S, Al-Saati Z N, Patryk K, Khaddar R A, Al-Jumeily D, Shaw A, Ruddock F and Aljefery M 2019 Assessment of heavy metal pollution in the Great Al-Mussaib irrigation channel *Desalination and Water Treatment* **168** 165-74

[8] Hashim K S, Al-Saati N H, Hussein A H and Al-Saati Z N 2018 An investigation into the level of heavy metals leaching from canal-dreged sediment: a case study metals leaching from dreged sediment. In: *First International Conference on Materials Engineering & Science,* (Istanbul Aydın University (IAU), Turkey pp 12-22

[9] Alenazi M, Hashim K S, Hassan A A, Muradov M, Kot P and Abdulhadi B 2020 Turbidity removal using natural coagulants derived from the seeds of strychnos potatorum: statistical and experimental approach. In: *IOP Conference Series: Materials Science and Engineering*: IOP Publishing) p 012064

[10] Hashim K S, Ali S S M, AlRifaie J K, Kot P, Shaw A, Al Khaddar R, Idowu I and Gkantou M 2020 Escherichia coli inactivation using a hybrid ultrasonic–electrocoagulation reactor *Chemosphere* **247** 125868-75

[11] Hashim K S, AlKhaddar R, Shaw A, Kot P, Al-Jumeily D, Alwash R and Aljefery M H 2020 *Electrocoagulation as an eco-friendly River water treatment method. In Advances in Water Resources Engineering and Management* (Berline: Springer)

[12] Abdulhadi B, Kot P, Hashim K, Shaw A, Muradov M and Al-Khaddar R 2021 Continuous-flow electrocoagulation (EC) process for iron removal from water: Experimental, statistical and economic study *Science of The Total Environment* **756** 1-16

[13] Hashim K S, Shaw A, AlKhaddar R, Kot P and Al-Shamma'a A 2021 Water purification from metal ions in the presence of organic matter using electromagnetic radiation-assisted treatment *Journal of Cleaner Production* **280**

[14] Hashim K S, Ewadh H M, Muhsin A A, Zubaidi S L, Kot P, Muradov M, Aljefery M and Al-Khaddar R 2020 Phosphate removal from water using bottom ash: Adsorption performance, coexisting anions and modelling studies *Water Science and Technology* **83** 1-17

[15] Hashim K S, Shaw A, Al Khaddar R, Ortoneda Pedrola M and Phipps D 2017 Defluoridation of drinking water using a new flow column-electrocoagulation reactor (FCER) - Experimental, statistical, and economic approach *Journal of Environmental Management* **197** 80-8

[16] Hashim K S, Shaw A, Al Khaddar R, Pedrola M O and Phipps D 2017 Energy efficient electrocoagulation using a new flow column reactor to remove nitrate from drinking water - Experimental, statistical, and economic approach *Journal of Environmental Management* **196** 224-33

[17] Alyafei A, AlKizwini R S, Hashim K S, Yeboah D, Gkantou M, Al Khaddar R, Al-Faluji D and Zubaidi S L 2020 Treatment of effluents of construction industry using a combined filtration-electrocoagulation method. In: *IOP Conference Series: Materials Science and Engineering*: IOP Publishing) p 012032

[18] Aqeel K, Mubarak H A, Amoako-Attah J, Abdul-Rahaim L A, Al Khaddar R, Abdellatif M, Al-Janabi A and Hashim K S 2020 Electrochemical removal of brilliant green dye from wastewater. In: *IOP Conference Series: Materials Science and Engineering*: IOP Publishing) p 012036

[19] Emamjomeh M M, Mousazadeh M, Mokhtari N, Jamali H A, Makkiabadi M, Naghdali Z, Hashim K S and Ghanbari R 2020 Simultaneous removal of phenol and linear alkylbenzene sulfonate from automotive service station wastewater: Optimization of coupled electrochemical and physical processes *Separation Science and Technology* **55** 3184-94

[20] Alenezi A K, Hasan H A, Hashim K S, Amoako-Attah J, Gkantou M, Muradov M, Kot P and Abdulhadi B 2020 Zeolite-assisted electrocoagulation for remediation of phosphate from calcium-phosphate solution. In: *IOP Conference Series: Materials Science and Engineering*: IOP Publishing) p 012031

[21] Alhendal M, Nasir M J, Hashim K S, Amoako-Attah J, Al-Faluji D, Muradov M, Kot P and Abdulhadi B 2020 Cost-effective hybrid filter for remediation of water from fluoride. In: *IOP Conference Series: Materials Science and Engineering*: IOP Publishing) p 012038

[22] Al-Marri S, AlQuzweeni S S, Hashim K S, AlKhaddar R, Kot P, AlKizwini R S, Zubaidi S L and Al-Khafaji Z S 2020 Ultrasonic-Electrocoagulation method for nitrate removal from water. In: *IOP Conference Series: Materials Science and Engineering*: IOP Publishing) p 012073

[23] Hashim K, Kot P, Zubaid S, Alwash R, Al Khaddar R, Shaw A, Al-Jumeily D and Aljefery M 2020 Energy efficient electrocoagulation using baffle-plates electrodes for efficient Escherichia Coli removal from Wastewater *Journal of Water Process Engineering* **33** 101079-86

[24] Alattabi A W, Harris C, Alkhaddar R, Alzeyadi A and Hashim K 2017 Treatment of Residential Complexes' Wastewater using Environmentally Friendly Technology *Procedia Engineering* **196** 792-9

[25] Alattabi A W, Harris C B, Alkhaddar R M, Hashim K S, Ortoneda-Pedrola M and Phipps D 2017 Improving sludge settleability by introducing an innovative, two-stage settling sequencing batch reactor *Journal of Water Process Engineering* **20** 207-16

[26] Hashim K S, Al-Saati N H, Alquzweeni S S, Zubaidi S L, Kot P, Kraidi L, Hussein A H, Alkhaddar R, Shaw A and Alwash R 2019 Decolourization of dye solutions by electrocoagulation: an investigation of the effect of operational parameters. In: *First International Conference on Civil and Environmental Engineering Technologies (ICCEET),* (University of Kufa, Iraq pp 25-32

[27] Hashim K S, Hussein A H, Zubaidi S L, Kot P, Kraidi L, Alkhaddar R, Shaw A and Alwash R 2019 Effect of initial pH value on the removal of reactive black dye from water by electrocoagulation (EC) method. In: *2nd International Scientific Conference,* (Al-Qadisiyah University, Iraq pp 12-22

[28] Hashim K S, Khaddar R A, Jasim N, Shaw A, Phipps D, Kot P, Pedrola M O, Alattabi A W, Abdulredha M and Alawsh R 2019 Electrocoagulation as a green technology for phosphate removal from River water *Separation and Purification Technology* **210** 135-44

[29] Abdulla G, Kareem M M, Hashim K S, Muradov M, Kot P, Mubarak H A, Abdellatif M and Abdulhadi B 2020 Removal of iron from wastewater using a hybrid filter. In: *IOP Conference Series: Materials Science and Engineering*: IOP Publishing) p 012035

[30] Abdulraheem F S, Al-Khafaji Z S, Hashim K S, Muradov M, Kot P and Shubbar A A 2020 Natural filtration unit for removal of heavy metals from water. In: *IOP Conference Series: Materials Science and Engineering*: IOP Publishing) p 012034

[31] Idowu I A, Atherton W, Hashim K, Kot P, Alkhaddar R, Alo B I and Shaw A 2019 An analyses of the status of landfill classification systems in developing countries: Sub Saharan Africa landfill experiences *Waste Management* **87** 761-71

[32] Abdulredha M, Al Khaddar R, Jordan D, Kot P, Abdulridha A and Hashim K 2018 Estimating solid waste generation by hospitality industry during major festivals: A quantification model based on multiple regression *Waste Management* **77** 388-400

[33] Abdulredha M, Kot P, Al Khaddar R, Jordan D and Abdulridha A 2020 Investigating municipal solid waste management system performance during the Arba'een event in the city of Kerbala, Iraq *Environment, Development and Sustainability* **22** 1431-54

[34] Abdulredha M, Rafid A, Jordan D and Hashim K 2017 The development of a waste management system in Kerbala during major pilgrimage events: determination of solid waste composition *Procedia Engineering* **196** 779-84

[35] Abdulhadi B A, Kot P, Hashim K S, Shaw A and Khaddar R A 2019 Influence of current density and electrodes spacing on reactive red 120 dye removal from dyed water using electrocoagulation/electroflotation (EC/EF) process. In: *First International Conference on Civil and Environmental Engineering Technologies (ICCEET),* (University of Kufa, Iraq pp 12-22

[36] Al-Jumeily D, Hashim K, Alkaddar R, Al-Tufaily M and Lunn J 2019 Sustainable and Environmental Friendly Ancient Reed Houses (Inspired by the Past to Motivate the Future). In: *11th International Conference on Developments in eSystems Engineering (DeSE),* (Cambridge, UK pp 214-9

[37] Hassan Alnaimi I J I, Abuduljaleel Al-Janabi, Khalid Hashim, Michaela Gkantou, Salah L. Zubaidi, Patryk Kot, Magomed Muradov 2020 Ultrasonic-electrochemical treatment for effluents of concrete plants Ultrasonic-electrochemical treatment for effluents of concrete plants. In: *IOP Conference Series Materials Science and Engineering* (University of Kufa, Najaf, Iraq pp 1-9

[38] Zanki A K, Mohammad F H, Hashim K S, Muradov M, Kot P, Kareem M M and Abdulhadi B 2020 Removal of organic matter from water using ultrasonic-assisted electrocoagulation method. In: *IOP Conference Series: Materials Science and Engineering*: IOP Publishing) p 012033

[39] Majdi H S, Shubbar A, Nasr M S, Al-Khafaji Z S, Jafer H, Abdulredha M, Masoodi Z A, Sadique M and Hashim K 2020 Experimental data on compressive strength and ultrasonic pulse velocity properties of sustainable mortar made with high content of GGBFS and CKD combinations *Data in Brief* **31** 105961-72

[40] Hashim K S, Idowu I A, Jasim N, Al Khaddar R, Shaw A, Phipps D, Kot P, Pedrola M O, Alattabi A W and Abdulredha M J M 2018 Removal of phosphate from River water using a new baffle plates electrochemical reactor **5** 1413-8

[41] Isra'a S S, Al-Janabi A, Abdulredha M, Alkandari A, Abdellatif M and Yeboah D 2021 Reusing of furnace bottom ash as an adsorbent for phosphate removal from water. In: *IOP Conference Series: Materials Science and Engineering*: IOP Publishing) p 012006

[42] Jawad S F, Saddam N S, Adaami Q J, Kareem M M, Abdulredha M, Mubarak H A, Kot P, Gkantou M and AlKhayyat A 2021 Dye removal from textile wastewater using solar-powered electrocoagulation reactor. In: *IOP Conference Series: Materials Science and Engineering*: IOP Publishing) p 012016

[43] Shubbar A A, Jafer H, Dulaimi A, Hashim K, Atherton W and Sadique M 2018 The development of a low carbon binder produced from the ternary blending of cement, ground granulated blast furnace slag and high calcium fly ash: An experimental and statistical approach *Construction and Building Materials* **187** 1051-60

[44] Shubbar A A, Al-Shaer A, AlKizwini R S, Hashim K, Hawesah H A and Sadique M 2019 Investigating the influence of cement replacement by high volume of GGBS and PFA on the mechanical performance of cement mortar. In: *First International Conference on Civil and Environmental Engineering Technologies (ICCEET),* (University of Kufa, Iraq pp 31-8

[45] Kadhim A, Sadique M, Al-Mufti R and Hashim K 2020 Long-term performance of novel high-calcium one-part alkali-activated cement developed from thermally activated lime kiln dust *Journal of Building Engineering* **32** 1-17
[46] Kadhim A, Sadique M, Al-Mufti R and Hashim K 2020 Developing One-Part Alkali-Activated metakaolin/natural pozzolan Binders using Lime Waste as activation Agent *Advances in Cement Research* **32** 1-38
[47] Shubbar A A, Sadique M, Nasr M S, Al-Khafaji Z S and Hashim K S 2020 The impact of grinding time on properties of cement mortar incorporated high volume waste paper sludge ash *Karbala International Journal of Modern Science* **6** 1-23
[48] Shubbar A A, Sadique M, Shanbara H K and Hashim K 2020 *The Development of a New Low Carbon Binder for Construction as an Alternative to Cement. In Advances in Sustainable Construction Materials and Geotechnical Engineering* (Berlin: Springer)
[49] Aayef A N, Al Masoodi W T M, Kamel R J, Abdulredha M, Almansoori N A, Kot P and Muradov M 2021 An experimental study for adapting electrocoagulation as a technique for fluoride removal from water. In: *IOP Conference Series: Materials Science and Engineering*: IOP Publishing) p 012012
[50] Abdulredha M, Abdulridha A, Shubbar A, Alkhaddar R, Kot P and Jordan D 2020 Estimating municipal solid waste generation from service processions during the Ashura religious event. In: *IOP Conference Series: Materials Science and Engineering*: IOP Publishing) p 012075
[51] Abdulredha M, Al Khaddar R and Jordan D 2017 Hoteliers' attitude towards solid waste source separation through mega festivals: A pilot study in Karbala. In: *International Conference for Doctoral Research*: BUID)
[52] Abdulredha M, Al Khaddar R, Jordan D, Al-Attabi A and Alzeyadi A 2017 Public participation in solid waste management during mega festivals: A pilot study. In: *WCST World Congress on Sustainable Technologies Proceedings 2017*: Infonomics Society) pp 38-41
[53] Abdulredha M, Al Khaddar R, Jordan D, Kot P, Abdulridha A and Hashim K J W M 2018 Estimating solid waste generation by hospitality industry during major festivals: A quantification model based on multiple regression **77** 388-400
[54] Abdulredha M, al-Khaddar R, Kot P, Jordan D and Abdulridha A 2018 Benchmarking of the Current Solid Waste Management System in Karbala, Iraq, Using Wasteaware Benchmark Indicators. In: *World Environmental and Water Resources Congress 2018: Groundwater, Sustainability, and Hydro-Climate/Climate Change*: American Society of Civil Engineers Reston, VA) pp 40-8
[55] Abdulredha M, Kadhim N, Hussein A, Almutairi M, Alkhaddar R, Yeboah D and Hashim K 2021 Zeolite as a natural adsorbent for nitrogenous compounds being removed from water. In: *IOP Conference Series: Materials Science and Engineering*: IOP Publishing) p 012082
[56] Abdulredha M, Kot P, Al Khaddar R, Jordan D, Abdulridha A J E, Development and Sustainability 2020 Investigating municipal solid waste management system performance during the Arba'een event in the city of Kerbala, Iraq **22** 1431-54
[57] Abdulredha M, Muhsin A A, Al-Janabi A, Alajmi B N, Gkantou M, Amoako-Attah J, Al-Jumeily D, Mustafina J and AlKhayyat A 2021 Using SF and CKD as cement replacement materials for producing cement mortar. In: *IOP Conference Series: Materials Science and Engineering*: IOP Publishing) p 012007
[58] Abdulredha M, Rafid A, Jordan D and Alattabi A J P e 2017 Facing up to waste: how can hotel managers in Kerbala, Iraq, help the city deal with its waste problem? **196** 771-8

[59] Asl F B, Leili M, Vaziri Y, Arian S S, Cristaldi A, Conti G O and Ferrante M 2018 Health impacts quantification of ambient air pollutants using AirQ model approach in Hamadan, Iran *Environmental research* **161** 114-21

[60] Shubbar A A, Jafer H, Abdulredha M, Al-Khafaji Z S, Nasr M S, Al Masoodi Z and Sadique M J J o B E 2020 Properties of cement mortar incorporated high volume fraction of GGBFS and CKD from 1 day to 550 days **30** 101327

[61] Majdi H S, Shubbar A, Nasr M S, Al-Khafaji Z S, Jafer H, Abdulredha M, Al Masoodi Z, Sadique M and Hashim K J D i B 2020 Experimental data on compressive strength and ultrasonic pulse velocity properties of sustainable mortar made with high content of GGBFS and CKD combinations

[62] Abdulredha M, Rafid A, Jordan D and Hashim K J P E 2017 The development of a waste management system in Kerbala during major pilgrimage events: determination of solid waste composition **196** 779-84

[63] Al-Anbari R, Alnakeeb A, Abdulredha M J E and Journal T 2013 Landfill site selection for Kerbala municipal solid wastes by using geographical information system techniques **32** 13

[64] Alattabi A W, Harris C, Alkhaddar R, Alzeyadi A and Abdulredha M J P e 2017 Online monitoring of a sequencing batch reactor treating domestic wastewater **196** 800-7

[65] Weichenthal S, Farrell W, Goldberg M, Joseph L and Hatzopoulou M 2014 Characterizing the impact of traffic and the built environment on near-road ultrafine particle and black carbon concentrations *Environmental research* **132** 305-10

Evaluation of the impact of municipal landfills on soil

S. Alganimi[1], **J. Alshibeep**[1*]
[1] College of Science, University of Kerbala, Iraq.

Email: mohammed@uokerbala.edu.iq

Abstract. Enormous quantities of municipal waste are produced around the world and the majority of that waste is disposed of into municipal landfill. These wastes have significant impacts on the environment surrounding them including soil. Many researchers focusing their attention to study the effects on municipal landfills on the surrounding environment. Therefore, the aim of this study was to determine the heavy metal pollution effects of a municipal solid waste landfill located in Kerbala, Iraq. The concentration of two heavy metal amounts that are chromium and nickel, were investigated at two sites near the landfill during 2020. The investigation sites were located about 15 and 30 meters from the edge of the targeted landfill. Disturbed soil samples were collected form the two sites a 2 meters depth. Then, the gathered samples were dried using in room temperature (20 ^{0}C) sieved through a 2 mm mesh. The levels of the selected heavy metals were calculated by means of plasma optical emission spectroscopy and linked to the standards of the environmental protection administration. It was found that the concentrations of both metals in close site (site 1) were the higher than the far site (site 2). The analyses also revealed that the concentration of the Nickle is higher than the concentration of the Chromium in both sites. Comparing the concentrations of the selected heavy metals in Kerbala' landfill with standards of the environmental protection administration, it was found that all of the assessed heavy metals were found to be within the set standards.

1. Introduction

The Heavy metals are an important part of the earth's ecosystem and naturally exist in fresh water, groundwater and soil[1-3]. According to research literature, iron, for example, can be found high levels up to 50 ppm in surface and ground water [4-7]. The presence of the heavy metals does not pose a direct danger to the human and eco-system since the environment will naturally keep the heavy metal concentrations at balance [8-11]. But, the cumulative impacts of people activity, have greatly changed the normal cycles of many heavy and other elements in the environment, making the eco-systems incapable of balancing the concentration of the heavy metal concentrations in the environment (water, air and soils)[12-15].

In general, the Heavy metals can come from a variety of sources and activated like transportation, agricultural, industrial, mining, planting etc. [16-22]. Heavy metals pollution is large linked with severs health impacts including Alzheimer, cancer, lung diseases, etc. [23-26], as well as harming the environment [13, 27, 28]. Recently, it has been reported that the impacts of heavy metals increased

significantly global warming on water supply [29-32]. Various treatment techniques like biological[33-37], chemical [33, 38, 39], electrochemical [34, 40-43] and biodegradation [35, 44-46], recovery technologies [30, 31, 36, 37] have been applied to strip heavy metals and other toxic contaminants from water. As the earth population is uncontrolledly expanding, the number and scale of sites used as landfills have risen significantly in order to handle the quantity of municipal solid waste directed to municipal landfills [19, 32, 47]. For instance, it was reported that the average annual production of solid waste in the urban areas of India is about 50 million tonnes. These huge quantities are predicted to be over 240 million tonnes by 2050. The same trend is observed in Iraq by researchers. Researchers revealed that solid waste generation is steadily increasing over the last two decades and the quantity of the solid waste is continually growing [1, 22, 48, 49]. Moreover, extreme quantities of demolishing solid waste are produced which is also disposed of in municipal landfills [2, 50-55]. In the city of Kerbala the quantity of solid waste disposed in landfills increased form 340 thousand tonnes per year during 2005 to reach about 700 thousand tonnes during 2018. These quantities are expected to be near 1 million tonnes per year during 2035. Additionally, Kerbala is affected by many events where the city is faced with instant increase in the quantity of solid waste due to the dramatic increase in the visitor population of the city. one example of large event is the Arba'een religious event where more than 18 million visitor enters the city over 15 days to do religious activities and in the same time produce large quantities of solid waste [3, 7, 19].
Unfortunately, most solid waste generated around the world is poorly managed. About 90 percent of the solid waste generated in India, for example, is poorly managed and disposed of without proper management unit [16, 18]. Solid waste disposal is only contaminating water bodies with a wide variety of contaminants, but also significantly contaminated the underlying soils with many types of pollutants, it has thus become an important issue to the green establishments. Furthermore, disposal sites produce many types of gasses that could lead to serious ecological hazards and higher water usage [26, 28, 56, 57].
Therefore, the current research investigation aims to study the impact of municipal landfill site located in the city of Kerbala on the surrounding soil. The concentration of two heavy metals that are chromium, and nickel in the areas surround the landfill site in order to understand the effects of the landfill on the surrounding soil.

2. Selected landfill site

The target landfill site in this investigation is situated in the southwest of the city of Kerbala (32° 32' 0" N and 32° 28' 0" N as shown in figure 1. This site collects various forms of urban solid waste, fluctuating from plain food refuse to building waste. Generally, the depth of the trenches used for waste disposal was calculated to be between 2 and 4 m at the disposal site. The landfill of Kerbala has been selected to examine the effects of municipal landfills on the surrounding soils because the landfill poorly maintained like the majority of landfills in Iraq. The waste is discarded lacking an appropriate disposal procedure. The landfill is regarded as dumping site without good engineering practice. Besides, the sit contains many rag pickers who market the recently discarded waste to the related businesses, who recycles bottles, canisters, metals and plastic wastes.

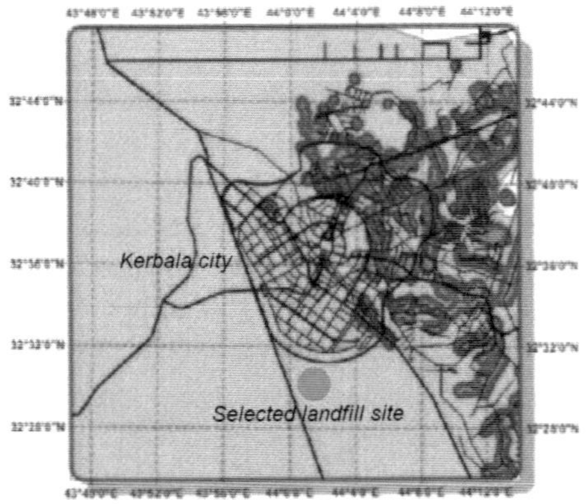

Figure 1: the location of the landfill site in the city of Kerbala.

3. Materials and methods

In this research, two soil samples were collected form two sites located about 15 and 30 meters form the landfill site. The soil samples were obtained from about 2 meters form the top level of the soil in each site. Plastic containers were uses to keep the samples in isolated space and preserve its characterise. The samples were transported to labs for analyses and examination in terms of heavy metals pollution.
At the lab, the related research literature was followed in terms of processing and preparation of the gathered samples [58]. Firstly, the obtained samples were dried out for one day at oven temperature of 65 °C. A 2 mm mesh was used for separating waste and stones/gravels to sew the dry samples. The seven samples were digested with inductively coupled plasma optical emission spectrometry before checking for Cr, Cu, Ni and Pb concentration (model: OPTIMA-2000). After drying, the samples were passed through 2 mm sieve to remove rubble and gravels/chippings. Finally, the samples were examined to find the consecration the chromium and nickel using optical spectroscopy.

4. Results and discussion

Two soil samples collected from two sites that are spread about 15 meters and 30 meters form the landfill site. Table 1 presents the concentration levels of the selected pollutants near the landfill site. The findings gotten shows that, as the distance from the landfill site increases, the concentration levels of the targeted heavy metals (chromium and nickel) are reduced. The connection between the distance from the selected landfill site and the reduction in the concentration of the heavy metal can be regarded from several aspects. One, there has been a drop in rainfall over recent years and a high drought that considerably restricted the penetration of surface rinse into the site and wash out the toxins into nearby land and water bodies. Besides, the water leaching from the accumulated waste in the dump site is normally move for very short distances. Thus, it can be seen that the concentration of the pollutant is increased near the landfill site and drops when the distance increases. Comparing the concertation of the heavy metals around the site with the standard limits, it is clearly the concentrations is below the limits which acceptable sign about the impact of the landfill on the surrounding soils [58].

Table 1: The concentrations level of the heavy metals near Kerbala landfill.

Pollutant	Site	Distance (m)	Concentration (mg/kg)	Average (mg/kg)	Limitations
Cr	1	15	37.7	31.1	250 mg/kg
	2	30	24.5		
Ni	1	15	46.1	38.5	60 mg/kg
	2	30	30.9		

5. Conclusions

The present research centred on investigating the effects of metal accumulation on the underlying soil within municipal landfill sites, concentrating mostly on chromium, and nickel levels. The findings revealed that the level of contaminants surveyed were negatively related the distance from the sites, at which the concentrations of chromium, and nickel was observed to be decreased as the location of the sampling site be distanced form the landfill site. In addition, the concertation level of the assed heavy metals were below guidelines prosed by the environmental protection. The Nickle have the highest concentration level in the sites investigated. Future researches are recommended to investigate the presence of other heavy metals loke cadmium and arsenic near the landfill site.

References

[1] Jawad S F, Saddam N S, Adaami Q J, Kareem M M, Abdulredha M, Mubarak H A, Kot P, Gkantou M and AlKhayyat A 2021 Dye removal from textile wastewater using solar-powered electrocoagulation reactor. In: *IOP Conference Series: Materials Science and Engineering*: IOP Publishing) p 012016

[2] Majdi H S, Shubbar A, Nasr M S, Al-Khafaji Z S, Jafer H, Abdulredha M, Al Masoodi Z, Sadique M and Hashim K J D i B 2020 Experimental data on compressive strength and ultrasonic pulse velocity properties of sustainable mortar made with high content of GGBFS and CKD combinations

[3] Shubbar A A, Jafer H, Abdulredha M, Al-Khafaji Z S, Nasr M S, Al Masoodi Z and Sadique M J J o B E 2020 Properties of cement mortar incorporated high volume fraction of GGBFS and CKD from 1 day to 550 days **30** 101327

[4] Abdulhadi B, Kot P, Hashim K, Shaw A, Muradov M and Al-Khaddar R 2021 Continuous-flow electrocoagulation (EC) process for iron removal from water: Experimental, statistical and economic study *Science of The Total Environment* **756** 1-16

[5] Hashim K S, Shaw A, AlKhaddar R, Kot P and Al-Shamma'a A 2021 Water purification from metal ions in the presence of organic matter using electromagnetic radiation-assisted treatment *Journal of Cleaner Production* **280**

[6] Abdulraheem F S, Al-Khafaji Z S, Hashim K S, Muradov M, Kot P and Shubbar A A 2020 Natural filtration unit for removal of heavy metals from water. In: *IOP Conference Series: Materials Science and Engineering*: IOP Publishing) p 012034

[7] Abdulredha M, Al Khaddar R and Jordan D 2017 Hoteliers' attitude towards solid waste source separation through mega festivals: A pilot study in Karbala. In: *International Conference for Doctoral Research*: BUID)

[8] Abdulla G, Kareem M M, Hashim K S, Muradov M, Kot P, Mubarak H A, Abdellatif M and Abdulhadi B 2020 Removal of iron from wastewater using a hybrid filter. In: *IOP Conference Series: Materials Science and Engineering*: IOP Publishing) p 012035

[9] Alhendal M, Nasir M J, Hashim K S, Amoako-Attah J, Al-Faluji D, Muradov M, Kot P and Abdulhadi B 2020 Cost-effective hybrid filter for remediation of water from fluoride. In: *IOP Conference Series: Materials Science and Engineering*: IOP Publishing) p 012038

[10] Al-Marri S, AlQuzweeni S S, Hashim K S, AlKhaddar R, Kot P, AlKizwini R S, Zubaidi S L and Al-Khafaji Z S 2020 Ultrasonic-Electrocoagulation method for nitrate removal from water. In: *IOP Conference Series: Materials Science and Engineering*: IOP Publishing) p 012073

[11] Abdulredha M, Al Khaddar R, Jordan D, Al-Attabi A and Alzeyadi A 2017 Public participation in solid waste management during mega festivals: A pilot study. In: *WCST World Congress on Sustainable Technologies Proceedings 2017*: Infonomics Society) pp 38-41

[12] Alyafei A, AlKizwini R S, Hashim K S, Yeboah D, Gkantou M, Al Khaddar R, Al-Faluji D and Zubaidi S L 2020 Treatment of effluents of construction industry using a combined filtration-electrocoagulation method. In: *IOP Conference Series: Materials Science and Engineering*: IOP Publishing) p 012032

[13] Aqeel K, Mubarak H A, Amoako-Attah J, Abdul-Rahaim L A, Al Khaddar R, Abdellatif M, Al-Janabi A and Hashim K S 2020 Electrochemical removal of brilliant green dye from wastewater. In: *IOP Conference Series: Materials Science and Engineering*: IOP Publishing) p 012036

[14] Emamjomeh M M, Mousazadeh M, Mokhtari N, Jamali H A, Makkiabadi M, Naghdali Z, Hashim K S and Ghanbari R 2020 Simultaneous removal of phenol and linear alkylbenzene sulfonate from automotive service station wastewater: Optimization of coupled electrochemical and physical processes *Separation Science and Technology* **55** 3184-94

[15] Abdulredha M, Al Khaddar R, Jordan D, Kot P, Abdulridha A and Hashim K J W M 2018 Estimating solid waste generation by hospitality industry during major festivals: A quantification model based on multiple regression **77** 388-400

[16] Idowu I A, Atherton W, Hashim K, Kot P, Alkhaddar R, Alo B I and Shaw A 2019 An analyses of the status of landfill classification systems in developing countries: Sub Saharan Africa landfill experiences *Waste Management* **87** 761-71

[17] Hashim K S, Shaw A, Al Khaddar R, Pedrola M O and Phipps D 2017 Iron removal, energy consumption and operating cost of electrocoagulation of drinking water using a new flow column reactor *Journal of Environmental Management* **189** 98-108

[18] Abdulredha M, Rafid A, Jordan D and Hashim K 2017 The development of a waste management system in Kerbala during major pilgrimage events: determination of solid waste composition *Procedia Engineering* **196** 779-84

[19] Abdulredha M, Al Khaddar R, Jordan D, Kot P, Abdulridha A and Hashim K 2018 Estimating solid waste generation by hospitality industry during major festivals: A quantification model based on multiple regression *Waste Management* **77** 388-400

[20] Hashim K S, Al-Saati N H, Hussein A H and Al-Saati Z N 2018 An investigation into the level of heavy metals leaching from canal-dreged sediment: a case study metals leaching from dreged sediment. In: *First International Conference on Materials Engineering & Science,* (Istanbul Aydın University (IAU), Turkey pp 12-22

[21] Hashim K S, Idowu I A, Jasim N, Al Khaddar R, Shaw A, Phipps D, Kot P, Pedrola M O, Alattabi A W and Abdulredha M 2018 Removal of phosphate from River water using a new baffle plates electrochemical reactor *MethodsX* **5** 1413-8

[22] Abdulredha M, al-Khaddar R, Kot P, Jordan D and Abdulridha A 2018 Benchmarking of the Current Solid Waste Management System in Karbala, Iraq, Using Wasteaware Benchmark Indicators. In: *World Environmental and Water Resources Congress 2018: Groundwater, Sustainability, and Hydro-Climate/Climate Change*: American Society of Civil Engineers Reston, VA) pp 40-8

[23] Hashim K S, Al-Saati N H, Alquzweeni S S, Zubaidi S L, Kot P, Kraidi L, Hussein A H, Alkhaddar R, Shaw A and Alwash R 2019 Decolourization of dye solutions by electrocoagulation: an investigation of the effect of operational parameters. In: *First International Conference on Civil and Environmental Engineering Technologies (ICCEET),* (University of Kufa, Iraq pp 25-32

[24] Hashim K S, Hussein A H, Zubaidi S L, Kot P, Kraidi L, Alkhaddar R, Shaw A and Alwash R 2019 Effect of initial pH value on the removal of reactive black dye from water by electrocoagulation (EC) method. In: *2nd International Scientific Conference,* (Al-Qadisiyah University, Iraq pp 12-22

[25] Hashim K S, Khaddar R A, Jasim N, Shaw A, Phipps D, Kot P, Pedrola M O, Alattabi A W, Abdulredha M and Alawsh R 2019 Electrocoagulation as a green technology for phosphate removal from River water *Separation and Purification Technology* **210** 135-44

[26] Abdulredha M, Kadhim N, Hussein A, Almutairi M, Alkhaddar R, Yeboah D and Hashim K 2021 Zeolite as a natural adsorbent for nitrogenous compounds being removed from water. In: *IOP Conference Series: Materials Science and Engineering*: IOP Publishing) p 012082

[27] Hashim K S, AlKhaddar R, Shaw A, Kot P, Al-Jumeily D, Alwash R and Aljefery M H 2020 *Electrocoagulation as an eco-friendly River water treatment method. In Advances in Water Resources Engineering and Management* (Berline: Springer)

[28] Abdulredha M, Kot P, Al Khaddar R, Jordan D, Abdulridha A J E, Development and Sustainability 2020 Investigating municipal solid waste management system performance during the Arba'een event in the city of Kerbala, Iraq **22** 1431-54

[29] Farhan S L, Jasim I A and Al-Mamoori S K 2019 The transformation of the city of Najaf, Iraq: Analysis, reality and future prospects *Journal of Urban Regeneration & Renewal* **13** 160-71

[30] Hashim K S, Al Khaddar R, Jasim N, Shaw A, Phipps D, Kot P, Pedrola M O, Alattabi A W, Abdulredha M, Alawsh R J S and Technology P 2019 Electrocoagulation as a green technology for phosphate removal from River water **210** 135-44

[31] Hashim K S, Idowu I A, Jasim N, Al Khaddar R, Shaw A, Phipps D, Kot P, Pedrola M O, Alattabi A W and Abdulredha M J M 2018 Removal of phosphate from River water using a new baffle plates electrochemical reactor **5** 1413-8

[32] Isra'a S S, Al-Janabi A, Abdulredha M, Alkandari A, Abdellatif M and Yeboah D 2021 Reusing of furnace bottom ash as an adsorbent for phosphate removal from water. In: *IOP Conference Series: Materials Science and Engineering*: IOP Publishing) p 012006

[33] Abdulredha M, Rafid A, Jordan D and Hashim K J P E 2017 The development of a waste management system in Kerbala during major pilgrimage events: determination of solid waste composition **196** 779-84

[34] Al-Anbari R, Alnakeeb A, Abdulredha M J E and Journal T 2013 Landfill site selection for Kerbala municipal solid wastes by using geographical information system techniques **32** 13

[35] Alattabi A W, Harris C, Alkhaddar R, Alzeyadi A and Abdulredha M J P e 2017 Online monitoring of a sequencing batch reactor treating domestic wastewater **196** 800-7

[36] Al-Sareji O J, Abdulredha M, Mubarak H A, Grmasha R A, Alnowaishry A, Kot P, Al-Khaddar R and AlKhayyat A 2021 Copper removal from water using carbonized sawdust. In: *IOP Conference Series: Materials Science and Engineering*: IOP Publishing) p 012015

[37] ALWAN H H, SALEH L A, AL-MOHAMMED F M, ABDULREDHA M A J J o E S and Technology 2020 EXPERIMENTAL PREDICTION OF THE DISCHARGE COEFFICIENTS FOR RECTANGULAR WEIR WITH BOTTOM ORIFICES **15** 3265-80

[38] Alenazi M, Hashim K S, Hassan A A, Muradov M, Kot P and Abdulhadi B 2020 Turbidity removal using natural coagulants derived from the seeds of strychnos potatorum: statistical and experimental approach. In: *IOP Conference Series: Materials Science and Engineering*: IOP Publishing) p 012064

[39] Alenezi A K, Hasan H A, Hashim K S, Amoako-Attah J, Gkantou M, Muradov M, Kot P and Abdulhadi B 2020 Zeolite-assisted electrocoagulation for remediation of phosphate from calcium-phosphate solution. In: *IOP Conference Series: Materials Science and Engineering*: IOP Publishing) p 012031

[40] Hashim K, Kot P, Zubaid S, Alwash R, Al Khaddar R, Shaw A, Al-Jumeily D and Aljefery M 2020 Energy efficient electrocoagulation using baffle-plates electrodes for efficient Escherichia Coli removal from Wastewater *Journal of Water Process Engineering* **33** 101079-86

[41] Hashim K S, Ali S S M, AlRifaie J K, Kot P, Shaw A, Al Khaddar R, Idowu I and Gkantou M 2020 Escherichia coli inactivation using a hybrid ultrasonic–electrocoagulation reactor *Chemosphere* **247** 125868-75

[42] Hashim K S, Shaw A, Al Khaddar R, Ortoneda Pedrola M and Phipps D 2017 Defluoridation of drinking water using a new flow column-electrocoagulation reactor (FCER) - Experimental, statistical, and economic approach *Journal of Environmental Management* **197** 80-8

[43] Hashim K S, Shaw A, Al Khaddar R, Pedrola M O and Phipps D 2017 Energy efficient electrocoagulation using a new flow column reactor to remove nitrate from drinking water -

Experimental, statistical, and economic approach *Journal of Environmental Management* **196** 224-33

[44] Alattabi A W, Harris C, Alkhaddar R, Alzeyadi A and Hashim K 2017 Treatment of Residential Complexes' Wastewater using Environmentally Friendly Technology *Procedia Engineering* **196** 792-9

[45] Alattabi A W, Harris C B, Alkhaddar R M, Hashim K S, Ortoneda-Pedrola M and Phipps D 2017 Improving sludge settleability by introducing an innovative, two-stage settling sequencing batch reactor *Journal of Water Process Engineering* **20** 207-16

[46] Hashim K S, Ewadh H M, Muhsin A A, Zubaidi S L, Kot P, Muradov M, Aljefery M and Al-Khaddar R 2020 Phosphate removal from water using bottom ash: Adsorption performance, coexisting anions and modelling studies *Water Science and Technology* **83** 1-17

[47] Abdulredha M, Abdulridha A, Shubbar A, Alkhaddar R, Kot P and Jordan D 2020 Estimating municipal solid waste generation from service processions during the Ashura religious event. In: *IOP Conference Series: Materials Science and Engineering*: IOP Publishing) p 012075

[48] Abdulredha M, Kot P, Al Khaddar R, Jordan D and Abdulridha A 2020 Investigating municipal solid waste management system performance during the Arba'een event in the city of Kerbala, Iraq *Environment, Development and Sustainability* **22** 1431-54

[49] Al-Jumeily D, Hashim K, Alkaddar R, Al-Tufaily M and Lunn J 2019 Sustainable and Environmental Friendly Ancient Reed Houses (Inspired by the Past to Motivate the Future). In: *11th International Conference on Developments in eSystems Engineering (DeSE),* (Cambridge, UK pp 214-9

[50] Shubbar A A, Al-Shaer A, AlKizwini R S, Hashim K, Hawesah H A and Sadique M 2019 Investigating the influence of cement replacement by high volume of GGBS and PFA on the mechanical performance of cement mortar. In: *First International Conference on Civil and Environmental Engineering Technologies (ICCEET),* (University of Kufa, Iraq pp 31-8

[51] Kadhim A, Sadique M, Al-Mufti R and Hashim K 2020 Long-term performance of novel high-calcium one-part alkali-activated cement developed from thermally activated lime kiln dust *Journal of Building Engineering* **32** 1-17

[52] Kadhim A, Sadique M, Al-Mufti R and Hashim K 2020 Developing One-Part Alkali-Activated metakaolin/natural pozzolan Binders using Lime Waste as activation Agent *Advances in Cement Research* **32** 1-38

[53] Majdi H S, Shubbar A, Nasr M S, Al-Khafaji Z S, Jafer H, Abdulredha M, Masoodi Z A, Sadique M and Hashim K 2020 Experimental data on compressive strength and ultrasonic pulse velocity properties of sustainable mortar made with high content of GGBFS and CKD combinations *Data in Brief* **31** 105961-72

[54] Shubbar A A, Sadique M, Nasr M S, Al-Khafaji Z S and Hashim K S 2020 The impact of grinding time on properties of cement mortar incorporated high volume waste paper sludge ash *Karbala International Journal of Modern Science* **6** 1-23

[55] Shubbar A A, Sadique M, Shanbara H K and Hashim K 2020 *The Development of a New Low Carbon Binder for Construction as an Alternative to Cement. In Advances in Sustainable Construction Materials and Geotechnical Engineering* (Berlin: Springer)

[56] Abdulredha M, Muhsin A A, Al-Janabi A, Alajmi B N, Gkantou M, Amoako-Attah J, Al-Jumeily D, Mustafina J and AlKhayyat A 2021 Using SF and CKD as cement replacement materials for producing cement mortar. In: *IOP Conference Series: Materials Science and Engineering*: IOP Publishing) p 012007

[57] Abdulredha M, Rafid A, Jordan D and Alattabi A J P e 2017 Facing up to waste: how can hotel managers in Kerbala, Iraq, help the city deal with its waste problem? **196** 771-8

[58] Khan S, Cao Q, Zheng Y, Huang Y and Zhu Y 2008 Health risks of heavy metals in contaminated soils and food crops irrigated with wastewater in Beijing, China *Environmental pollution* **152** 686-92

Producing cement mortars through implementing cement replacement

J. Mohammed[1], R. Ali[1,*]
[1] Civil Engineer, Ministry of Construction, Iraq.
Email: heshamkhaliq@gmail.com

Abstract. Cement mortars have been recognised as one of the most important and effective binding materials. The main materials that can be used in cement mortars are sand and Portland cement. The implementation of natural materials in the production process of cement mortars has many negative environmental effects including energy consumption production of various pollutants. Scientific research highlighted that the implementation of wasted materials as cementitious materials has large potential environmentally and economically. The present research study aims to assess the compressive strength of cement mortars that contain replacement Cementous materials that are cement kiln dust and silica fume. The compressive strength of the developed mortars was examined after 7 and 28 days of curing. The findings extracted from the specimens that contain cement replacement have been compared to the specimens of cement mortar having only cement. The findings revealed that mixtures that contain cement kiln dust and silica fume have less compressive strength than mixtures of ordinary Portland cement only. Moreover, a lesser compressive strength was obtained with a greater cement kiln dust content. Samples with reduced replacement at various curing times have higher compressive strength values.

Keywords: Cement kiln dust, compressive strength. silica fume, cement mortars.

1. Introduction

In the past, water, cement and sand can be considered as the constituents of concrete. The mortars of cement have wide applications in buildings and the construction sector around the world due to the high working strength and easy application [1, 2]. Cement mortars have been largely implemented as a joining part and even fill the spaces in construction locations, particularly in-residence places and minor projects. Additionally, there are many other applications of the cement mortar like epoxy materials for bridging construction parts [3, 4].

Raw resources like sand and cement formed by grinding limestone are implemented in the production of Portland cement. Many legislation concerning the uses of such raw resources and potentially harmful environmental and economic effects have recently been created [5, 6]. Along with the failure problems of land segments which can occur with increased utilization of raw resources, the environmental damage from raw materials consumption may be described as a significantly higher level of contamination. Moreover, the chemical properties of concrete cause CO_2 to be released, thus increasing power usage and the detrimental effects on the atmosphere[7-11]. Carbon dioxide is accountable for numerous climate impacts, including such potable water scarcities [12-15] and irregular global diversion of water resources[16-19]. Furthermore, cement-based industries wastewater is extremely corrosive and retains a high amount of nitrogen material and solid materials [20-22] which damage the environment the watercourses received and thus effective handling procedures are required to reduce this contamination [23-26]. The most common treatment methods used to reduce wastewater contamination in many methods include electrocoagulation [27-36], electrocoagulation [27-36], adsorption [37-40], natural coagulants [41, 42], advanced methods [43] wand even biotic units [44, 45].

In another extreme, increased manufacturing and maintaining prices, with increased demand for raw resources, will contribute the adverse financial effects [46, 47]. The removal of concrete buildings is yet another problem in addition to such results since these refuses should be deposited and treated at protected sites requiring the massive costs involved [48, 49]. Both the above issues with the use of raw resources are contrary to renewable technology principles [50]. A further argument is that more homes have to be constructed, which would also demand further building resources until green substitutes are created [51, 52]. Owing to such significant growth, advanced inventions that maintain low costs and appropriate load-bearing efficiency are used to substitute raw resources. Alternate solution items used are mostly wastes, such as industrial refuse and agricultural wastes, to help improve the atmosphere and reduce costs as low as possible [53, 54]. Agriculture produces thousands of tonnes of various wasted material like seed, weeds and roots root vegetables, oil palm petrol, maize policeman, and tobacco, per year which can be used in the construction sector [48, 55]. Such refuse amounts without adequate monitoring or preparation resulted in different impacts for people and the planet. In several situations, the disposal of refused materials is distributed to vast areas adding to greater concerns as area farmland, construction as well as other social concerns continually is reduced. The gathered waste is typically treated through the combustion of excess smoke and CO2 emissions in the landfill zone. Thus, it is important to analyse the potential to use such wastes for various industries [3, 56, 57].

Various waste forms, like wood ash, base-ash, limestone, fuel ash, etc. have been examined as the substitute for cement, which produced a great value and suitable characteristics. Mortar of cement is known as an adapted blender of cementitious or base metals as a cement substitute. The addition of cementitious substance as a partial substitution for cement may improve mortar's bearing capacity and longevity [58-60]. Silica fume is produced in the thermal power station as a refused waste product. The silica fume has also been used as a cement substitute for a wide variety of plants to increase the mortars consistency in terms of compression efficiency and longevity. The kiln dust is among the major waste generated by the processing facilities. The texture (finely ground substance) and chemical properties of the product are similar to the Portland cement. One of the negative aspects of kiln dust is the absence of cementitious products, which could be overcome by mixing other pozzolanic with its toughness and resistance property as a limited Portland cement substitute. This analysis explores the characteristics of cement mortar having kiln dust and silica fume used as cement substitutes.

2. Materials and methods

Several mixtures that contain various percentages of Portland cement substitution of kiln dust and silica fume were prepared using several percentages of the replacement materials. The percentages of the replacement materials used in this research project are presented in figures 1. From the figure, it can be seen that five mixtures were developed each mixture contains various percentages of replacement materials. In mixture 0, Portland cement only was used as cementitious materials for comparison purposes. Then, various percentages of the kiln dust and silica fume were used t replace the cement in the other 4 mixtures. These percentages were used to examine the effects of waste materials on the compressive strength of the developed mortars.

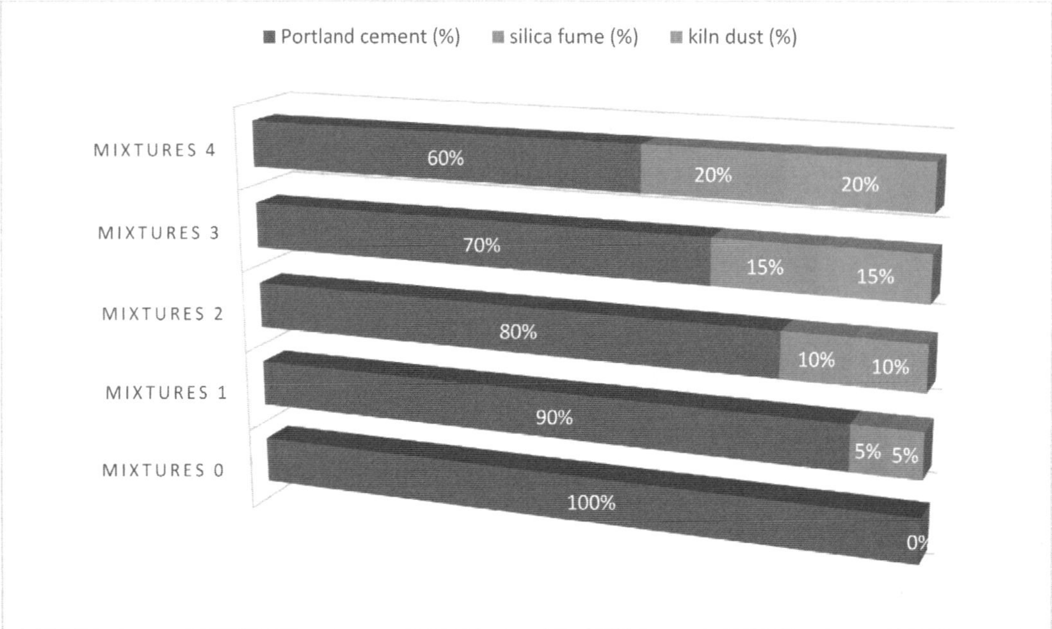

Figure 1: Mortars' mixtures developed in this research.

2.1. Compressive strength test

The British Standard (BS EN 196-1) was followed to conduct the sample preparation and testing to evaluate the compressive strength of the developed mortars using various percentages of waste materials namely silica fume and kiln dust. Initially, dry materials were correctly blended by implementing an electronic blender to ensure homogenisation [61]. The obtained mixture is then combined with the right water proportion. Several prisms that have dimensions of 40mm * 40mm * 160mm were filled with the samples. In order to facilitate the removal after hardening, each prism was greased until the mortar was placed. The specimens were packaged in plastic bags to reduce water loss during ageing. Three samples were prepared and tested of each mixture are to ensure the trustworthiness of the information. Then, the developed samples were tested at age 7 days and 28 days to understand the development of the compressive strength of the mortars that implement silica fume and kiln dust.

3. Results and discussion

The test of the compressive strength of the mortar's findings indicates that perhaps the compressive strength of Portland cement was only the best among all of 28 treatment days based on the results presented in figure 2. The compressive strength of the mortars developed from cement only without any waste materials was 9.5 MPa after 7 days of curing. Then, after 14 days of curing the compressive strength increased from 9.5 MPa to about 10.2 MPa. The compressive strength of the developed mortars using cement only reached 11.1 after 28 days of curing. These samples were produced and tested to compare the mortars that developed using percentages of waste materials with the compressive strength using cement only.

The second group of mortars was developed using only 10% of waste materials (5% kiln dust and 5% silica fume). from figure 2, it can be said that the compressive strength of the mortars containing only 10 waste materials dropped by about 8% among all curing ages. When cement replacement increased from 10% to 20%, the compressive strength of the developed mortars with cement replacement, even more, dropped to only about 85% of the compressive strength of mortars without cement replacements. From the figure, it can be seen that the compressive strength reaches only 9.3 MPa after 28 days of curing. The compressive strength continues to drop with the increase in the percentage of the replacement materials to reach only 60% after adding 20% of Klin dust and 20% of the silica fume. The findings of this research are in agreement with the conclusions of Shoaib and Balaha [62]. Shoaib and Balaha [62] reported that the compressive strength is reduced with the increase in the fraction of kiln dust replacement.

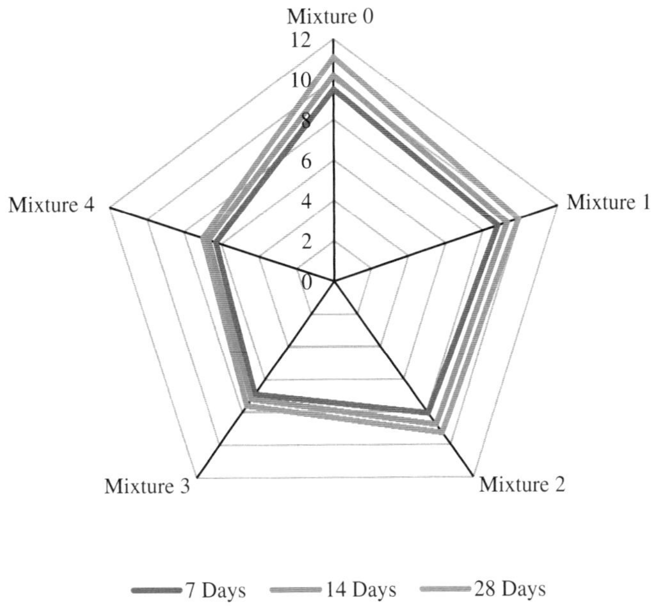

Figure 2: The compressive strength of the developed mixtures using various percentages of kiln dust and silica fume.

4. Conclusions

The usage of wasted materials as cement replacement has been wildly investigated. This research analyses the application of kiln dust and silica fume to replace Portland cement in the production of cement mortars. Various percentages were used to assess analyses the impact of waste materials on the development of the compressive strength. Based on the results, it can be concluded that mortars compressive strength has a negative relationship with the percentage of the replaced cementitious materials.

References

[1] Abdulredha M, al-Khaddar R, Kot P, Jordan D and Abdulridha A 2018 Benchmarking of the Current Solid Waste Management System in Karbala, Iraq, Using Wasteaware Benchmark Indicators. In: *World Environmental and Water Resources Congress 2018: Groundwater, Sustainability, and Hydro-Climate/Climate Change*: American Society of Civil Engineers Reston, VA) pp 40-8

[2] Abdulredha M, Kadhim N, Hussein A, Almutairi M, Alkhaddar R, Yeboah D and Hashim K 2021 Zeolite as a natural adsorbent for nitrogenous compounds being removed from water. In: *IOP Conference Series: Materials Science and Engineering*: IOP Publishing) p 012082

[3] Shubbar A A, Sadique M, Shanbara H K and Hashim K 2020 *The Development of a New Low Carbon Binder for Construction as an Alternative to Cement. In Advances in Sustainable Construction Materials and Geotechnical Engineering* (Berlin: Springer)

[4] Shubbar A A, Jafer H, Dulaimi A, Hashim K, Atherton W and Sadique M 2018 The development of a low carbon binder produced from the ternary blending of cement, ground

granulated blast furnace slag and high calcium fly ash: An experimental and statistical approach *Construction and Building Materials* **187** 1051-60
[5] Naqi A and Jang J G 2019 Recent progress in green cement technology utilizing low-carbon emission fuels and raw materials: A review *Sustainability* **11** 537
[6] Shubbar A A, Al-Shaer A, AlKizwini R S, Hashim K, Hawesah H A and Sadique M 2019 Investigating the influence of cement replacement by high volume of GGBS and PFA on the mechanical performance of cement mortar. In: *First International Conference on Civil and Environmental Engineering Technologies (ICCEET),* (University of Kufa, Iraq pp 31-8
[7] Aayef A N, Al Masoodi W T M, Kamel R J, Abdulredha M, Almansoori N A, Kot P and Muradov M 2021 An experimental study for adapting electrocoagulation as a technique for fluoride removal from water. In: *IOP Conference Series: Materials Science and Engineering*: IOP Publishing) p 012012
[8] Abdulredha M, Abdulridha A, Shubbar A, Alkhaddar R, Kot P and Jordan D 2020 Estimating municipal solid waste generation from service processions during the Ashura religious event. In: *IOP Conference Series: Materials Science and Engineering*: IOP Publishing) p 012075
[9] Abdulredha M, Al Khaddar R and Jordan D 2017 Hoteliers' attitude towards solid waste source separation through mega festivals: A pilot study in Karbala. In: *International Conference for Doctoral Research*: BUID)
[10] Abdulredha M, Al Khaddar R, Jordan D, Al-Attabi A and Alzeyadi A 2017 Public participation in solid waste management during mega festivals: A pilot study. In: *WCST World Congress on Sustainable Technologies Proceedings 2017*: Infonomics Society) pp 38-41
[11] Abdulredha M, Al Khaddar R, Jordan D, Kot P, Abdulridha A and Hashim K J W M 2018 Estimating solid waste generation by hospitality industry during major festivals: A quantification model based on multiple regression **77** 388-400
[12] Abdulredha M, Kot P, Al Khaddar R, Jordan D, Abdulridha A J E, Development and Sustainability 2020 Investigating municipal solid waste management system performance during the Arba'een event in the city of Kerbala, Iraq **22** 1431-54
[13] Abdulredha M, Muhsin A A, Al-Janabi A, Alajmi B N, Gkantou M, Amoako-Attah J, Al-Jumeily D, Mustafina J and AlKhayyat A 2021 Using SF and CKD as cement replacement materials for producing cement mortar. In: *IOP Conference Series: Materials Science and Engineering*: IOP Publishing) p 012007
[14] Abdulredha M, Rafid A, Jordan D and Alattabi A J P e 2017 Facing up to waste: how can hotel managers in Kerbala, Iraq, help the city deal with its waste problem? **196** 771-8
[15] Abdulredha M, Rafid A, Jordan D and Hashim K J P E 2017 The development of a waste management system in Kerbala during major pilgrimage events: determination of solid waste composition **196** 779-84
[16] Al-Anbari R, Alnakeeb A, Abdulredha M J E and Journal T 2013 Landfill site selection for Kerbala municipal solid wastes by using geographical information system techniques **32** 13
[17] Alattabi A W, Harris C, Alkhaddar R, Alzeyadi A and Abdulredha M J P e 2017 Online monitoring of a sequencing batch reactor treating domestic wastewater **196** 800-7
[18] Al-Sareji O J, Abdulredha M, Mubarak H A, Grmasha R A, Alnowaishry A, Kot P, Al-Khaddar R and AlKhayyat A 2021 Copper removal from water using carbonized sawdust. In: *IOP Conference Series: Materials Science and Engineering*: IOP Publishing) p 012015
[19] ALWAN H H, SALEH L A, AL-MOHAMMED F M, ABDULREDHA M A J J o E S and Technology 2020 EXPERIMENTAL PREDICTION OF THE DISCHARGE COEFFICIENTS FOR RECTANGULAR WEIR WITH BOTTOM ORIFICES **15** 3265-80
[20] Alyafei A, AlKizwini R S, Hashim K S, Yeboah D, Gkantou M, Al Khaddar R, Al-Faluji D and Zubaidi S L 2020 Treatment of effluents of construction industry using a combined filtration-electrocoagulation method. In: *IOP Conference Series: Materials Science and Engineering*: IOP Publishing) p 012032

[21] Hassan Alnaimi I J I, Abuduljaleel Al-Janabi, Khalid Hashim, Michaela Gkantou, Salah L. Zubaidi, Patryk Kot, Magomed Muradov 2020 Ultrasonic-electrochemical treatment for effluents of concrete plants Ultrasonic-electrochemical treatment for effluents of concrete plants. In: *IOP Conference Series Materials Science and Engineering* (University of Kufa, Najaf, Iraq pp 1-9

[22] Zanki A K, Mohammad F H, Hashim K S, Muradov M, Kot P, Kareem M M and Abdulhadi B 2020 Removal of organic matter from water using ultrasonic-assisted electrocoagulation method. In: *IOP Conference Series: Materials Science and Engineering*: IOP Publishing) p 012033

[23] Hashim K S, Al-Saati N H, Alquzweeni S S, Zubaidi S L, Kot P, Kraidi L, Hussein A H, Alkhaddar R, Shaw A and Alwash R 2019 Decolourization of dye solutions by electrocoagulation: an investigation of the effect of operational parameters. In: *First International Conference on Civil and Environmental Engineering Technologies (ICCEET),* (University of Kufa, Iraq pp 25-32

[24] Hashim K S, Hussein A H, Zubaidi S L, Kot P, Kraidi L, Alkhaddar R, Shaw A and Alwash R 2019 Effect of initial pH value on the removal of reactive black dye from water by electrocoagulation (EC) method. In: *2nd International Scientific Conference,* (Al-Qadisiyah University, Iraq pp 12-22

[25] Omran I I, Al-Saati N H, Hashim K S, Al-Saati Z N, Patryk K, Khaddar R A, Al-Jumeily D, Shaw A, Ruddock F and Aljefery M 2019 Assessment of heavy metal pollution in the Great Al-Mussaib irrigation channel *Desalination and Water Treatment* **168** 165-74

[26] Hashim K S, Al-Saati N H, Hussein A H and Al-Saati Z N 2018 An investigation into the level of heavy metals leaching from canal-dreged sediment: a case study metals leaching from dreged sediment. In: *First International Conference on Materials Engineering & Science,* (Istanbul Aydın University (IAU), Turkey pp 12-22

[27] Aqeel K, Mubarak H A, Amoako-Attah J, Abdul-Rahaim L A, Al Khaddar R, Abdellatif M, Al-Janabi A and Hashim K S 2020 Electrochemical removal of brilliant green dye from wastewater. In: *IOP Conference Series: Materials Science and Engineering*: IOP Publishing) p 012036

[28] Hashim K, Kot P, Zubaid S, Alwash R, Al Khaddar R, Shaw A, Al-Jumeily D and Aljefery M 2020 Energy efficient electrocoagulation using baffle-plates electrodes for efficient Escherichia Coli removal from Wastewater *Journal of Water Process Engineering* **33** 101079-86

[29] Hashim K S, Ali S S M, AlRifaie J K, Kot P, Shaw A, Al Khaddar R, Idowu I and Gkantou M 2020 Escherichia coli inactivation using a hybrid ultrasonic–electrocoagulation reactor *Chemosphere* **247** 125868-75

[30] Mohammed A-H, Hussein A H, Yeboah D, Al Khaddar R, Abdulhadi B, Shubbar A A and Hashim K S 2020 Electrochemical removal of nitrate from wastewater. In: *IOP Conference Series: Materials Science and Engineering*: IOP Publishing) p 012037

[31] Hashim K S, Shaw A, AlKhaddar R, Kot P and Al-Shamma'a A 2021 Water purification from metal ions in the presence of organic matter using electromagnetic radiation-assisted treatment *Journal of Cleaner Production* **280**

[32] Hashim K S, Shaw A, Al Khaddar R, Ortoneda Pedrola M and Phipps D 2017 Defluoridation of drinking water using a new flow column-electrocoagulation reactor (FCER) - Experimental, statistical, and economic approach *Journal of Environmental Management* **197** 80-8

[33] Hashim K S, Shaw A, Al Khaddar R, Pedrola M O and Phipps D 2017 Iron removal, energy consumption and operating cost of electrocoagulation of drinking water using a new flow column reactor *Journal of Environmental Management* **189** 98-108

[34] Hashim K S, Shaw A, Al Khaddar R, Pedrola M O and Phipps D 2017 Energy efficient electrocoagulation using a new flow column reactor to remove nitrate from drinking water - Experimental, statistical, and economic approach *Journal of Environmental Management* **196** 224-33

[35] Abdulhadi B A, Kot P, Hashim K S, Shaw A and Khaddar R A 2019 Influence of current density and electrodes spacing on reactive red 120 dye removal from dyed water using electrocoagulation/electroflotation (EC/EF) process. In: *First International Conference on Civil and Environmental Engineering Technologies (ICCEET),* (University of Kufa, Iraq pp 12-22

[36] Hashim K S, Khaddar R A, Jasim N, Shaw A, Phipps D, Kot P, Pedrola M O, Alattabi A W, Abdulredha M and Alawsh R 2019 Electrocoagulation as a green technology for phosphate removal from River water *Separation and Purification Technology* **210** 135-44

[37] Abdulla G, Kareem M M, Hashim K S, Muradov M, Kot P, Mubarak H A, Abdellatif M and Abdulhadi B 2020 Removal of iron from wastewater using a hybrid filter. In: *IOP Conference Series: Materials Science and Engineering*: IOP Publishing) p 012035

[38] Abdulraheem F S, Al-Khafaji Z S, Hashim K S, Muradov M, Kot P and Shubbar A A 2020 Natural filtration unit for removal of heavy metals from water. In: *IOP Conference Series: Materials Science and Engineering*: IOP Publishing) p 012034

[39] Alenezi A K, Hasan H A, Hashim K S, Amoako-Attah J, Gkantou M, Muradov M, Kot P and Abdulhadi B 2020 Zeolite-assisted electrocoagulation for remediation of phosphate from calcium-phosphate solution. In: *IOP Conference Series: Materials Science and Engineering*: IOP Publishing) p 012031

[40] Hashim K S, Ewadh H M, Muhsin A A, Zubaidi S L, Kot P, Muradov M, Aljefery M and Al-Khaddar R 2020 Phosphate removal from water using bottom ash: Adsorption performance, coexisting anions and modelling studies *Water Science and Technology* **83** 1-17

[41] Al-Jumeily D, Hashim K, Alkaddar R, Al-Tufaily M and Lunn J 2019 Sustainable and Environmental Friendly Ancient Reed Houses (Inspired by the Past to Motivate the Future). In: *11th International Conference on Developments in eSystems Engineering (DeSE),* (Cambridge, UK pp 214-9

[42] Al-Saati N H, Hussein T K, Abbas M H, Hashim K, Al-Saati Z N, Kot P, Sadique M, Aljefery M H and Carnacina I 2019 Statistical modelling of turbidity removal applied to non-toxic natural coagulants in water treatment: a case study *Desalination and Water Treatment* **150** 406-12

[43] Al-Marri S, AlQuzweeni S S, Hashim K S, AlKhaddar R, Kot P, AlKizwini R S, Zubaidi S L and Al-Khafaji Z S 2020 Ultrasonic-Electrocoagulation method for nitrate removal from water. In: *IOP Conference Series: Materials Science and Engineering*: IOP Publishing) p 012073

[44] Alattabi A W, Harris C, Alkhaddar R, Alzeyadi A and Hashim K 2017 Treatment of Residential Complexes' Wastewater using Environmentally Friendly Technology *Procedia Engineering* **196** 792-9

[45] Alattabi A W, Harris C B, Alkhaddar R M, Hashim K S, Ortoneda-Pedrola M and Phipps D 2017 Improving sludge settleability by introducing an innovative, two-stage settling sequencing batch reactor *Journal of Water Process Engineering* **20** 207-16

[46] Shubbar A A, Sadique M, Nasr M S, Al-Khafaji Z S and Hashim K S 2020 The impact of grinding time on properties of cement mortar incorporated high volume waste paper sludge ash *Karbala International Journal of Modern Science* **6** 1-23

[47] Kadhim A, Sadique M, Al-Mufti R and Hashim K 2020 Long-term performance of novel high-calcium one-part alkali-activated cement developed from thermally activated lime kiln dust *Journal of Building Engineering* **32** 1-17

[48] Abdulredha M, Rafid A, Jordan D and Hashim K 2017 The development of a waste management system in Kerbala during major pilgrimage events: determination of solid waste composition *Procedia Engineering* **196** 779-84

[49] Idowu I A, Atherton W, Hashim K, Kot P, Alkhaddar R, Alo B I and Shaw A 2019 An analyses of the status of landfill classification systems in developing countries: Sub Saharan Africa landfill experiences *Waste Management* **87** 761-71

[50] Kadhim A, Sadique M, Al-Mufti R and Hashim K 2020 Developing One-Part Alkali-Activated metakaolin/natural pozzolan Binders using Lime Waste as activation Agent *Advances in Cement Research* **32** 1-38

[51] Shubbar A A, Sadique M, Kot P and Atherton W 2019 Future of clay-based construction materials–A review *Construction and Building Materials* **210** 172-87

[52] Majdi H S, Shubbar A, Nasr M S, Al-Khafaji Z S, Jafer H, Abdulredha M, Masoodi Z A, Sadique M and Hashim K 2020 Experimental data on compressive strength and ultrasonic pulse velocity properties of sustainable mortar made with high content of GGBFS and CKD combinations *Data in Brief* **31** 105961-72

[53] Majdi H S, Shubbar A, Nasr M S, Al-Khafaji Z S, Jafer H, Abdulredha M, Al Masoodi Z, Sadique M and Hashim K J D i B 2020 Experimental data on compressive strength and ultrasonic pulse velocity properties of sustainable mortar made with high content of GGBFS and CKD combinations

[54] Shubbar A A, Jafer H, Abdulredha M, Al-Khafaji Z S, Nasr M S, Al Masoodi Z and Sadique M J J o B E 2020 Properties of cement mortar incorporated high volume fraction of GGBFS and CKD from 1 day to 550 days **30** 101327

[55] Abdulredha M, Al Khaddar R, Jordan D, Kot P, Abdulridha A and Hashim K 2018 Estimating solid waste generation by hospitality industry during major festivals: A quantification model based on multiple regression *Waste Management* **77** 388-400

[56] Farhan S L, Jasim I A and Al-Mamoori S K 2019 The transformation of the city of Najaf, Iraq: Analysis, reality and future prospects *Journal of Urban Regeneration & Renewal* **13** 160-71

[57] Hashim K S, Al Khaddar R, Jasim N, Shaw A, Phipps D, Kot P, Pedrola M O, Alattabi A W, Abdulredha M, Alawsh R J S and Technology P 2019 Electrocoagulation as a green technology for phosphate removal from River water **210** 135-44

[58] Hashim K S, Idowu I A, Jasim N, Al Khaddar R, Shaw A, Phipps D, Kot P, Pedrola M O, Alattabi A W and Abdulredha M J M 2018 Removal of phosphate from River water using a new baffle plates electrochemical reactor **5** 1413-8

[59] Isra'a S S, Al-Janabi A, Abdulredha M, Alkandari A, Abdellatif M and Yeboah D 2021 Reusing of furnace bottom ash as an adsorbent for phosphate removal from water. In: *IOP Conference Series: Materials Science and Engineering*: IOP Publishing) p 012006

[60] Jawad S F, Saddam N S, Adaami Q J, Kareem M M, Abdulredha M, Mubarak H A, Kot P, Gkantou M and AlKhayyat A 2021 Dye removal from textile wastewater using solar-powered electrocoagulation reactor. In: *IOP Conference Series: Materials Science and Engineering*: IOP Publishing) p 012016

[61] Shubbar A A, Jafer H, Abdulredha M, Al-Khafaji Z S, Nasr M S, Al Masoodi Z and Sadique M 2020 Properties of cement mortar incorporated high volume fraction of GGBFS and CKD from 1 day to 550 days *Journal of Building Engineering* 101327

[62] Shoaib M and Balaha M 2004 THERMO-CHEMICAL STABILITY AND MECHANICAL PROPERTIES OF MORTAR MADE WITH CEMENT KILN DUST-BLENDED CEMENT *ERJ. Engineering Research Journal* **27** 49-58

A short review on dyes removal from water and wastewaters

Hayfaa A. Mubarak[1,*], Marwa A. Kubba[2], K. Hashim[3], Abduljaleel Al-Janabi[4], Safaa K.H[4]

[1] Chemical Engineering Department, University of Babylon, Babel, Iraq
[2] Al-Rasheed University College, Baghdad, Iraq
[3] College of Environment, Al-Qasim Green University, Babylon, Iraq
[4] Engineering Faculty, University of Babylon, Babel, Iraq

[*] E-mail: haifaadnan_81@uobabylon.edu.iq

Abstract. Dyes are applied broadly in the textile industry to colour the products and give the products the final desire look. Additionally, dyes and colouring agents are used in a wide range of industries, like the papers production, packing, food and plastic industries. This means dyes and pigments represents an important pile in the global economy and civilisation. However, dyes used in this wide range of applications has a grave impact on human life due to the serious effects of the wasted dyes, in wastewater, on aquatic life, the chemistry of water and human health. Hence, the need for the removal of dyes of the produced wastewaters is an important step in any type of industry to avoid the mentioned impacts. The removal of dyes or colouring agents could be achieved in several ways, such as adsorption on active surfaces, electrolysing the colouring agent, application of oxidation chemicals, and filtration of coloured solution. This paper shortly reviews the removal of dyes from solutions by electro-coagulation method because this method is common nowadays, especially in developing countries. The results of this short review highlighted the electro-coagulation method is an effective and affordable method and suitable for the removal of dyes from water or wastewater. The EC method is also safe for the environment because there are no chemical additives in the treatment. Thus, this method could be a good option for developing countries.

1. Introduction

Dyes or colouring agents are substances or materials that chemically bond to the materials exposed to, and it is different from other chemicals that do not form chemical bonds with dyed or coloured materials, the latter called pigments [1]. The dyes could be classified according to:

1.1. Their sources

a) Natural dyes: produced from natural materials, such as berries, jack fruits, tea leaves, turmeric, fungi, indigo, and animals
b) Synthetic dyes: produced by mixing two or more chemicals, such as Alcian blue, Sulpho orange, Cresyl blue and reactive red

1.2. Their chemical structures

a) Disperse dye, such as disperse yellow 3
b) Reactive dye, such as reactive dye 5
c) Solvent Dye, such as solvent yellow 32

IOP Conf. Series: Materials Science and Engineering 1184 (2021) 012017 doi:10.1088/1757-899X/1184/1/012017

d) Direct dye, such as direct red 12B
e) Acidic dye, such as red acid 8
f) Basic Dyes, such as Acid yellow 36
g) Vat Dyes, such as Vat Black 25

Generally, the dyeing process is done by adding the dye or the agent into water or other aqueous solutions, and then the produced solution will be applied to the products that could be fabrics, food products, packing materials or other materials [2-4].

The dyeing process is not a new human activity, where dyed flax fibres (dated to 36,000 BP) were discovered in a prehistoric cave in Georgia. In fact, the dying process was used during the Neolithic periods, when the people started dying their fabrics natural materials such as fruits, peels, animals and roots and leaves of plants, for example, the crimson kermes was produced during the ancient and medieval worlds, and it was considered as a luxurious type of dyes. Additionally, dyes that were produced using woods, indigo and saffron were found in ancient Asia and Europe. The good fact about natural dyes is their minor effects on the environment and public health [5-7].

Unfortunately, the production of natural dyes is not enough to meet the increasing demand for dyed products, such as fabrics, backing items and food, due to the patterns of modern life [7]. For example, the global population has increased by many folds during the last few decades [8-10], and the urban areas were expanded by many folds [11-15]. As a result, the need for dyed clothes [16], food [17], packing products [18] and other daily needs [19-22] has increased rapidly and hugely; therefore, synthetic dyes were produced and became the predominant type of dyes nowadays in all industries [23-25]. The first type of synthetic dyes, called mauve, was developed serendipitously by William H. Perkin in 1856, followed by a dramatic increase in the production and consumption of dyes and colouring agents, especially during and after the 21st century. Nowadays, the number of the synthetic dyes is more than 100,000 types, and the global production of these dyes is more than 700, 000 tons/year. The average consumption of freshwater during the dyeing process is between 100 and 200 litters per 1 kg of fabrics depending on the colouring technology and type of fabrics.

This increase in the consumption of dyes was accompanied by a significant increase in the generation of dyed wastewaters that has the main role in the deterioration of water quality and the spread of serious diseases [26-28]. For instance, dyes could cause carcinogenic diseases due to their chemical compositions that contain toxic elements, besides the consumption of the dissolved oxygen in the aquatic environment, the limitation of the sunlight penetration in surface water, and poising the aquatic life [29-31].

Therefore, several treatment approaches were experienced to remove the residual dyes and colouring agents from wastewaters before discharging them to the sources of water, aerobic and anaerobic degradations [32], adsorptions [33], filtrations [34], and electrochemical treatments [35-37] are good examples of the nowadays used approaches for the removal of dyes from wastewaters. Not all mentioned methods are applicable in poor or developing countries because many advanced methods, such as membrane filtration, require high capital cost and well-trained staff to operate them, which can not be found in many developing countries [38-40]. Therefore, many developing countries focus on the use of locally available methods, which do not require high cots or trained staff to operate, such as the electro-coagulation method.

The Electrocoagulation method is adopted on a large scale in developing countries because it can be manufactured using affordable local materials. However, this method is also operated by not well-trained staff. Figure 1 shows the general EC system. Therefore, this paper highlights the concepts of the EC method and some of the recent studies about the use of this method in the treatment of water and wastewater.

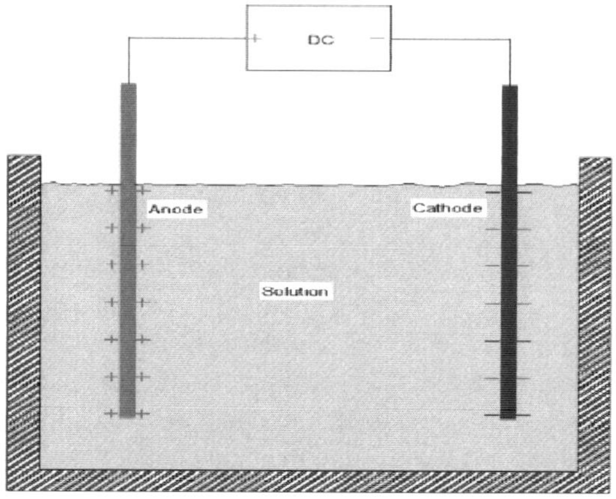

Figure 1: The EC system.

2. Electro-coagulation method

2.1. Principles of the EC method

Electrocoagulation (EC) method is one of the electrochemical methods that depend on metallic electrodes to produce, in situ, coagulants to remove the pollutants [3].

Generally, the EC units are made from anode and cathode (could be more than one pair) connected to a DC power. The DC current motivates the anode to produce positively charged ions and the cathode to produce hydrogen gas. The ions travel towards the negative electrode (the cathode); during this travel, the negative ions react with the pollutants in the solution (they usually have a positive charge), forming a floc (Figure 2), the latter grow in size until it reaches a heavyweight that can not be carried by the solution anymore [1]. As a result, the heavy flocs will be removed from the solution via precipitation and collected from the bottom of the unit as sludge. Another removal scenario happens when the applied current is high that motivate the cathode to produce more hydrogen gas in the form of bubbles. Because

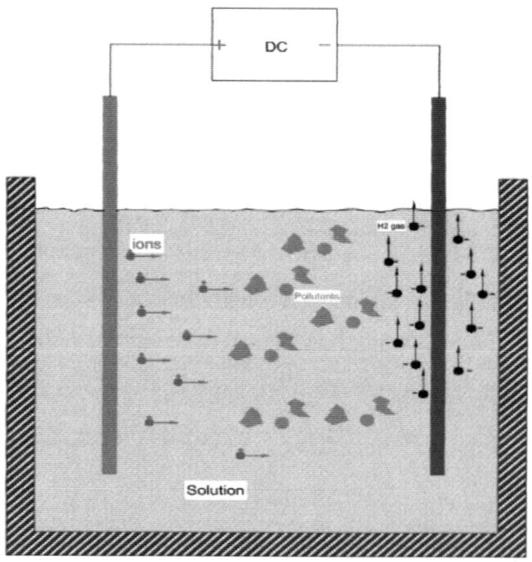

Figure 2: The mechanisms of the EC method.

the produced bubbles are lighter than the solution, they will move upward to leave the solution; during this travel, flocs will be attached to these bubbles and accumulated on the surface of the solution as foam, which will be scammed later using a metallic plate [5].

2.2. *The reaction in the EC method*

The reactions in the EC method depends on the type of electrodes. Because the most used metals in the EC system is iron and aluminium. Therefore, the reactions of the aluminium and iron are listed here as follows [23]:

2.2.1. *Aluminium electrodes*

At anodes:

$$Al(s) \rightarrow Al^{3+} + 3e^- \quad \textbf{(1)}$$

At cathodes:

$$2H_2O + 2e^- \rightarrow H_2(gas) + 2OH^- \quad \textbf{(2)}$$

Then, Al speciated according to the following equations:

$$Al^{3+} + H_2O \rightarrow Al(OH)^{2+} + H^+ \quad \textbf{(3)}$$

$$Al(OH)^{2+} + H_2O \rightarrow Al(OH)_2^+ + H^+ \quad \textbf{(4)}$$

$$Al(OH)_2^+ + H_2O \rightarrow Al(OH)_3^0 + H^+ \quad \textbf{(5)}$$

$$Al(OH)_3^0 + H_2O \rightarrow Al(OH)_{4\,(aq)}^- \quad \textbf{(6)}$$

2.2.2. *Iron electrodes*

The following reactions take place when iron (Fe) electrodes are used [41]:

At anodes:

$$4Fe(s) \rightarrow 4Fe^2 + 8e^- \quad \textbf{(7)}$$

$$4Fe^2 + 10H_2O + O_2 \rightarrow 4Fe(OH)_3 + 8H^+ \quad \textbf{(8)}$$

At cathodes:

$$8H^+ + 8e^- \rightarrow 4Fe(OH)_3 + 4H_2 \quad \textbf{(9)}$$

The overall reaction is:

$$4Fe + 10H_2O + O_2 \rightarrow 4Fe(OH)_3 + 4H_2 \quad \textbf{(10)}$$

2.3. *Advantages of the EC method*

The electro-coagulation method acquired wide attention because of many advantages, which can be summarised as follows [42-46]:

1- The flocs are generated because of the magnetic fields (positive and negative charges); therefore, there is no water captured inside the flocs, which means the expensive dewatering process. Thus, this feature makes the EC cost-effective method.

2- The EC method is run using DC power, which could be obtained from the sunlight, wind or dry batteries; therefore, the EC method could be portable and used in disasters

3- The EC is easy to be operated without the need for trained staff

4- The EC uses metallic electrodes to produce coagulants; therefore, no need to use chemicals in the treatment. Hence, the EC has no negative impacts on the environment

These advantages are behind the wide use of the EC in the current time, especially in developing countries.

However, some minor drawbacks of the EC method are the appearance of the inert layers on the aluminium electrodes and the lack of designs of the EC units.

2.4. Recent uses of the EC method in the treatment of water and wastewater

Table 1 summarises some of the recent studies about the use of the EC in the treatment of both water and wastewater.

Table 1: Some of the recent studies about the use of the EC in the treatment of both water and wastewater

Dye	Metal (electrodes)	Flow	Operating parameters (optimum)	Best removal	Authors
Acid Orange II	Granular activated carbon	Batch	Time = 30 min, cell voltage = 20V, air flow = 0.1m3/L, gap between electrodes = 14 mm.	87%	[47]
Acid Red 14	Iron anode and steel cathode.	Batch	Time = 4 min, current density = 8 mA/cm^2, pH = 6-9, gap between electrodes = 4 mm.	83%	Daneshvar, Sorkhabi and Kasiri [48]
Eosin yellowish	Mild steel	Batch	Time = 15 min, current density = 1.61 mA/cm^2, Power consumption = 1.5 KWh/m^3, pH = 6.8, gap between electrodes = 12 mm.	97%	Golder, Hridaya, Samanta and Ray [49]
Acid yellow 36	Iron	Batch	Time = 6 min, current density = 12.78 mA/cm^2, pH = 8, gap between electrodes = 25 mm.	83%	Kashefialasl, Khosravi, Marandi and Seyyedi [50]
Levafix orange E3 GA	Aluminium	Batch	Time = 12 min, current density = 10 mA/cm^2, Power consumption = 35	95%	Kobya, Demirbas, Can and

			kWh/kg dye, pH = 6.4, gap between electrodes = 11 mm.		Bayramoglu [51]
Red dye (2-naphthoic acid and 2-naphtol)	Aluminium	Continuous	Time = 14 min, current density = 31.25 mA/cm^2, Power consumption = 3.2 kWh/kg dye, flow rate = 37.2 L/h, pH = 6-9, gap between electrodes = 10 mm.	85%	Merzouk, Gourich, Sekki, Madani, Vial and Barkaoui [52]
Reactive Black 5	Iron	Batch	Time = 5 min, current density = 4.575 mA/cm^2, Power consumption = 4.96 kWh/kg dye, pH = 5, gap between electrodes = 25 mm.	98.8%	Sengil and Ozacar [53]
Orange II	Aluminium	continuous	Time = 7.3 min, current density = 16 mA/cm^2, Power consumption = 3.2 kWh/kg dye, pH = 6.5, flow rate = 350 mL/min, gap between electrodes = 6 mm.	94.5%	Mollah, Gomes, Das and Cocke [54]
Direct red 81	Aluminium	Batch	Time = 60 min, current density = 1.875 mA/cm^2, pH = 6, gap between electrodes = 15 mm,	98%	Aoudj, Khelifa, Drouiche, Hecini and Hamitouche [55]
Reactive Red198	Aluminium	Batch	Time = 30 min, cell voltage = 20V, pH = 5.5, Power consumption = 1.303 kW/m^3, gap between electrodes = 10 mm,	98.6%	Dalvand, Gholami, Joneidi and Mahmoodi [56]
Acid Yellow 220	Aluminium	Batch	Time = 7.5 min, current density = 4 mA/cm^2, pH = 5, Power consumption = 950 kWh/m^3, gap between electrodes = 10 mm.	97%	Pajootan, Arami and Mahmoodi [57]
Direct red 81	Aluminium	Continues	Time = 7.3 min, current density = 20 mA/cm^2, Power consumption =	90.2%	Zodi, Merzouk, Potier,

			3.2 kWh/kg dye, pH = 7.5, flow rate = 10 L/H, gap between electrodes = 10 mm.		Lapicque and Leclerc [58]
Methylene blue	Iron	Batch	Time = 24 min, current density = 5 mA/cm^2, Power consumption = 3.8 kWh/m^3 , pH = 9, gap between electrodes = 10 mm.	100%	Alizadeh, Ghahramani , Zarrabi and Hashemi [59]
Imperon violet KB	Aluminium	Batch	Time = 10 min, current density = 4 mA/cm^2, Power consumption = 4.66 kWh/m^3 , pH = 4.57, gap between electrodes = 10 mm.	98.5%	Naje, Chelliapan, Zakaria and Abbas [60]
Orange II dye	Aluminium	Batch	Current density =2.5A, stirring speed=450 rpm treatment time=280 minutes, and treatment cost of 4723-16629 US dollars	More than 99%	Signorelli, Costa and de Almeida Neto [61]
Acid orange 2	Iron electrodes		pH of solution=7, the current density=2 mA/cm^2, gap between electrodes=0.5 cm, and treatment time=40 minutes.	100%	Jawad, Saddam, Adaami, Kareem, Abdulredha, Mubarak, Kot, Gkantou and AlKhayyat [62]

3. Conclusion

The obtained results from this short review showed many facts about the EC method, which are:

1- The EC method is eco-friendly as it does depend on metallic plates to provide the coagulants instead of the additions of external chemicals; this means no secondary pollutants will be produced.
2- The operating cost of the EC method is low because it depends on the DC current that could be provided using sunlight, winds or dry batteries.
3- The flocs are generated because of the magnetic fields (positive and negative charges); therefore, there is no water captured inside the flocs, which means the expensive dewatering process. Thus, this feature makes the EC cost-effective method.
4- The EC is easy to be operated without the need for skilled staff.

5- Although the EC has many advantages, it has some drawbacks, such as the appearance of the inert layers on the aluminium electrodes and the lack of designs of the EC units.

References

[1] Hashim K S, Al-Saati N H, Alquzweeni S S, Zubaidi S L, Kot P, Kraidi L, Hussein A H, Alkhaddar R, Shaw A and Alwash R 2019. Decolourization of dye solutions by electro-coagulation: an investigation of the effect of operational parameters. *First International Conference on Civil and Environmental Engineering Technologies,* University of Kufa, Iraq.

[2] Safaa K, Shaw A, Al Khaddar R, Pedrola M O and Phipps D 2017. Iron removal, energy consumption and operating cost of electro-coagulation of drinking water using a new flow column reactor. *Journal of Environmental Management*, **189** 98-108.

[3] Khalid S, Hussein A H, Zubaidi S L, Kot P, Kraidi L, Alkhaddar R, Shaw A and Alwash R 2019. Effect of initial pH value on the removal of reactive black dye from water by electro-coagulation (EC) method. *2nd International Scientific Conference,* Al-Qadisiyah University, Iraq 12-22.

[4] Hashim K S, Shaw A, Al Khaddar R, Ortoneda Pedrola M and Phipps D 2017. Defluoridation of drinking water using a new flow column-electrocoagulation reactor (FCER) - Experimental, statistical, and economic approach. *Journal of Environmental Management*, **197** 80-8.

[5] Abdulhadi B A, Kot P, Hashim K S, Shaw A and Khaddar R A 2019. Influence of current density and electrodes spacing on reactive red 120 dye removal from dyed water using electro-coagulation/electroflotation (EC/EF) process. *First International Conference on Civil and Environmental Engineering Technologies (ICCEET),* University of Kufa, Iraq 12-22.

[6] Al-Saati N H, Hussein T K, Abbas M H, Al-Saati Z N, Kot P, Sadique M, Aljefery M H and Carnacina I 2019. Statistical modelling of turbidity removal applied to non-toxic natural coagulants in water treatment: a case study. *Desalination and Water Treatment*, **150** 406-12.

[7] Khaddar R A, Jasim N, Shaw A, Phipps D, Kot P, Pedrola M O, Alattabi A W, Abdulredha M and Alawsh R 2019. Electrocoagulation as a green technology for phosphate removal from River water. *Separation and Purification Technology*, **210** 135-44.

[8] Zubaidi S, Al-Bugharbee H, Muhsin Y R, Hashim K and Alkhaddar R 2020. Forecasting of monthly stochastic signal of urban water demand: Baghdad as a case study. *IOP Conference Series: Materials Science and Engineering,* 012018.

[9] Zubaidi S, Ortega-Martorell S, Al-Bugharbee H, Olier I, Hashim K S, Gharghan S K, Kot P and Al-Khaddar R 2020. Urban Water Demand Prediction for a City that Suffers from Climate Change and Population Growth: Gauteng Province case study. *Water*, **12** 1-18.

[10] Zanki A K, Mohammad F H, Hashim K S, Muradov M, Kot P, Kareem M M and Abdulhadi B 2020. Removal of organic matter from water using ultrasonic-assisted electro-coagulation method. *IOP Conference Series: Materials Science and Engineering,* 012033.

[11] Al-Jumeily D, Alkaddar R, Al-Tufaily M and Lunn J 2019. Sustainable and Environmental Friendly Ancient Reed Houses (Inspired by the Past to Motivate the Future). *11th International Conference on Developments in eSystems Engineering (DeSE),* Cambridge, UK 214-9.

[12] Farhan S L, Hashim I A J and Naji A A 2019. The Sustainable House: Comparative Analysis of Houses in Al Kut Neighborhoods-Iraq. *2019 12th International Conference on Developments in eSystems Engineering (DeSE),* 1031-6.

[13] Al-Saati N H, Omran I I, Salman A A, Al-Saati Z and Hashim K S 2021. Statistical modeling of monthly streamflow using time series and artificial neural network models: Hindiya Barrage as a case study. *Water Practice and Technology*, **16** 681-91.

[14] Al-Sareji O J, Grmasha R A, Salman J M, Idowu I and Hashim K S 2021. Street dust contamination by heavy metals in Babylon governorate, Iraq. *Journal of Engineering Science and Technology*, **16** 3528 - 46.

[15] Farhan S L, Antón D, Akef V S, Zubaidi S L and Hashim K S 2021. Factors influencing the transformation of Iraqi holy cities: the case of Al-Najaf. *Scientific Review Engineering and Environmental Sciences*, **30** 365-75.

[16] Abdulhadi B, Kot P, Hashim K, Shaw A, Muradov M and Al-Khaddar R 2021. Continuous-flow electro-coagulation (EC) process for iron removal from water: Experimental, statistical and economic study. *Science of The Total Environment*, **760** 1-16.
[17] Emamjomeh M M, Mousazadeh M, Mokhtari N, Jamali H A, Makkiabadi M, Naghdali Z, Hashim K S and Ghanbari R 2020. Simultaneous removal of phenol and linear alkylbenzene sulfonate from automotive service station wastewater: Optimisation of coupled electrochemical and physical processes. *Separation Science and Technology*, **55** 3184-94.
[18] Mousazadeh M, Paital B, Naghdali Z, Mortezania Z, Hashemi M, Karamati Niaragh E, Aghababaei M, Ghorbankhani M, Lichtfouse E, Sillanpää M, Hashim K S and Emamjomeh M M 2021. Positive environmental effects of the coronavirus 2020 episode: a review. *Environment, Development and Sustainability*, **21** 1-23.
[19] Hashim K S, Shaw A, AlKhaddar R, Kot P and Al-Shamma'a A 2021. Water purification from metal ions in the presence of organic matter using electromagnetic radiation-assisted treatment. *Journal of Cleaner Production*, **280** 1-17.
[20] Idowu I A, Hashim K, Shaw A and Nunes L J 2021. Enhancing the fuel properties of beverage wastes as non-edible feedstock for biofuel production. *Biofuels*, **14** 1-8.
[21] Abdulredha M, Rafid A, Jordan D and Hashim K 2017. The development of a waste management system in Kerbala during major pilgrimage events: determination of solid waste composition. *Procedia Engineering*, **196** 779-84.
[22] Alattabi A W, Harris C, Alzeyadi A and Hashim K 2017. Treatment of Residential Complexes' Wastewater using Environmentally Friendly Technology. *Procedia Engineering*, **196** 792-9.
[23] Hashim K S, Shaw A, Pedrola M O and Phipps D 2017. Energy efficient electro-coagulation using a new flow column reactor to remove nitrate from drinking water - Experimental, statistical, and economic approach. *Journal of Environmental Management*, **196** 224-33.
[24] Hashim K S, Al-Saati N H, Hussein A H and Al-Saati Z N 2018. An investigation into the level of heavy metals leaching from canal-dreged sediment: a case study metals leaching from dreged sediment. *First International Conference on Materials Engineering & Science,* Istanbul Aydın University (IAU), Turkey 12-22.
[25] Hashim K S, Idowu I A, Jasim N, Al Khaddar R, Shaw A, Phipps D, Kot P, Pedrola M O, Alattabi A W and Abdulredha M 2018. Removal of phosphate from River water using a new baffle plates electrochemical reactor. *MethodsX*, **5** 1413-8.
[26] Abdulla G, Kareem M M, Hashim K S, Muradov M, Kot P, Mubarak H A, Abdellatif M and Abdulhadi B 2020. Removal of iron from wastewater using a hybrid filter. *IOP Conference Series: Materials Science and Engineering,* 012035.
[27] Abdulraheem F S, Al-Khafaji Z S, Hashim K S, Muradov M, Kot P and Shubbar A A 2020. Natural filtration unit for removal of heavy metals from water. *IOP Conference Series: Materials Science and Engineering,* 012034.
[28] Alenazi M, Hashim K S, Hassan A A, Muradov M, Kot P and Abdulhadi B 2020. Turbidity removal using natural coagulants derived from the seeds of strychnos potatorum: statistical and experimental approach. *IOP Conference Series: Materials Science and Engineering,* 012064.
[29] Alenezi A K, Hasan H A, Hashim K S, Amoako-Attah J, Gkantou M, Muradov M, Kot P and Abdulhadi B 2020. Zeolite-assisted electro-coagulation for remediation of phosphate from calcium-phosphate solution. *IOP Conference Series: Materials Science and Engineering,* 012031.
[30] Alhendal M, Nasir M J, Hashim K S, Amoako-Attah J, Al-Faluji D, Muradov M, Kot P and Abdulhadi B 2020. Cost-effective hybrid filter for remediation of water from fluoride. *IOP Conference Series: Materials Science and Engineering,* 012038.
[31] Al-Mansori N J H, Alfatlawi T J M, Hashim K S and Al-Zubaidi L S 2020. The Effects of Different Shaped Baffle Blocks on the Energy Dissipation. *Civil Engineering Journal*, **6** 961-73.
[32] Al-Marri S, AlQuzweeni S S, Hashim K S, AlKhaddar R, Kot P, AlKizwini R S, Zubaidi S L and Al-Khafaji Z S 2020. Ultrasonic-Electrocoagulation method for nitrate removal from water. *IOP Conference Series: Materials Science and Engineering,* 012073.

[33] Alnaimi H, Idan I J, Al-Janabi A, Hashim K, Gkantou M, Zubaidi S L, Kot P and Muradov M 2020. Ultrasonic-electrochemical treatment for effluents of concrete plants. **888** 1-9.
[34] Alyafei A, AlKizwini R S, Hashim K S, Yeboah D, Gkantou M, Al Khaddar R, Al-Faluji D and Zubaidi S L 2020. Treatment of effluents of construction industry using a combined filtration-electrocoagulation method. *IOP Conference Series: Materials Science and Engineering,* 012032.
[35] Aqeel K, Mubarak H A, Amoako-Attah J, Abdul-Rahaim L A, Al Khaddar R, Abdellatif M, Al-Janabi A and Hashim K S 2020. Electrochemical removal of brilliant green dye from wastewater. *IOP Conference Series: Materials Science and Engineering,* 012036.
[36] Emamjomeh M M, Kakavand S, Jamali H A, Alizadeh S M, Safdari M, Mousavi S E S, Hashim K S and Mousazade M 2020. The treatment of printing and packaging wastewater by electro-coagulation–flotation: the simultaneous efficacy of critical parameters and economics. *Desalination and water treatment*, **205** 161-74.
[37] Grmasha R A, Al-sareji O J, Salman J M, Hashim K S and Jasim I A 2020. Polycyclic Aromatic Hydrocarbons (PAHs) in Urban Street Dust Within Three Land-Uses of Babylon Governorate, Iraq: Distribution, Sources, and Health Risk Assessment. *Journal of King Saud University - Engineering Sciences*, **33** 1-18.
[38] Hashim K, Kot P, Zubaid S, Alwash R, Al Khaddar R, Shaw A, Al-Jumeily D and Aljefery M 2020. Energy efficient electro-coagulation using baffle-plates electrodes for efficient Escherichia Coli removal from wastewater. *Journal of Water Process Engineering*, **33** 101079-86.
[39] Hashim K S, Ali S S M, AlRifaie J K, Kot P, Shaw A, Al Khaddar R, Idowu I and Gkantou M 2020. Escherichia coli inactivation using a hybrid ultrasonic–electro-coagulation reactor. *Chemosphere*, **247** 125868-75.
[40] Hashim K S, AlKhaddar R, Shaw A, Kot P, Al-Jumeily D, Alwash R and Aljefery M H 2020 *Electrocoagulation as an eco-friendly River water treatment method. In Advances in Water Resources Engineering and Management*, Berline: Springer.
[41] Nanseu-Njiki C P, Tchamango S R, Ngom P C, Darchen A and Ngameni E 2009. Mercury (II) removal from water by electro-coagulation using aluminium and iron electrodes. *Journal of Hazardous Materials*, **168** 1430-6.
[42] Hashim K S 2017 The innovative use of electrocoagulation-microwave techniques for the removal of pollutants from water. PhD thesis, *Civil Engineering,* Liverpool John Moores University, United Kingdom, p 229
[43] Hashim K, Shaw A and Phipps D, 2019. *Treatment reactor and method of treating a liquid.* Patent, WIPO, PCT/GB2019/052493, LJM University, United Kingdom.
[44] Omran I I, Al-Saati N H, Hashim K S, Al-Saati Z N, Patryk K, Khaddar R A, Al-Jumeily D, Shaw A, Ruddock F and Aljefery M 2019. Assessment of heavy metal pollution in the Great Al-Mussaib irrigation channel. *Desalination and Water Treatment*, **168** 165-74.
[45] Shubbar A A, Al-Shaer A, AlKizwini R S, Hashim K, Hawesah H A and Sadique M 2019. Investigating the influence of cement replacement by high volume of GGBS and PFA on the mechanical performance of cement mortar. *First International Conference on Civil and Environmental Engineering Technologies (ICCEET),* University of Kufa, Iraq 31-8.
[46] Hashim K S, Ewadh H M, Muhsin A A, Zubaidi S L, Kot P, Muradov M, Aljefery M and Al-Khaddar R 2020. Phosphate removal from water using bottom ash: Adsorption performance, coexisting anions and modelling studies. *Water Science and Technology*, **83** 1-17.
[47] Xiong Y, Strunk P j, Xia H, Zhu X and Karlsson H t 2001. Treatment of dye wastewater containing acid orange II using a cell with three-phase three-dimensional electrode. *Water research*, **35** 4226–30.
[48] Daneshvar N, Sorkhabi H A and Kasiri M B 2004. Decolorisation of dye solution containing Acid Red 14 by electro-coagulation with a comparative investigation of different electrode connections. *J Hazard Mater*, **112** 55-62.
[49] Golder A K, Hridaya N, Samanta A N and Ray S 2005. Electro-coagulation of methylene blue and eosin yellowish using mild steel electrodes. *J Hazard Mater*, **127** 134-40.

[50] Kashefialasl M, Khosravi M, Marandi R and Seyyedi K 2006. Treatment of dye solution containing colored index acid yellow 36 by electro-coagulation using iron electrodes. *International Journal of Environmental Science and Technology*, **2** 365-71.
[51] Kobya M, Demirbas E, Can O T and Bayramoglu M 2006. Treatment of levafix orange textile dye solution by electro-coagulation. *J Hazard Mater*, **132** 183-8.
[52] Merzouk B, Gourich B, Sekki A, Madani K, Vial C and Barkaoui M 2009. Studies on the decolorisation of textile dye wastewater by continuous electro-coagulation process. *Chemical Engineering Journal*, **149** 207-14.
[53] Sengil I A and Ozacar M 2009. The decolorisation of CI Reactive Black 5 in aqueous solution by electro-coagulation using sacrificial iron electrodes. *J Hazard Mater*, **161** 1369-76.
[54] Mollah M, Gomes J, Das K and Cocke D 2010. Electrochemical treatment of Orange II dye solution--use of aluminum sacrificial electrodes and floc characterisation. *J Hazard Mater*, **174** 851-8.
[55] Aoudj S, Khelifa A, Drouiche N, Hecini M and Hamitouche H 2010. Electro-coagulation process applied to wastewater containing dyes from textile industry. *Chemical Engineering and Processing: Process Intensification*, **49** 1176-82.
[56] Dalvand A, Gholami M, Joneidi A and Mahmoodi N M 2011. Dye Removal, Energy Consumption and Operating Cost of Electrocoagulation of Textile Wastewater as a Clean Process. *CLEAN - Soil, Air, Water*, **39** 665-72.
[57] Pajootan E, Arami M and Mahmoodi N M 2012. Binary system dye removal by electro-coagulation from synthetic and real colored wastewaters. *Journal of the Taiwan Institute of Chemical Engineers*, **43** 282-90.
[58] Zodi S, Merzouk B, Potier O, Lapicque F and Leclerc J-P 2013. Direct red 81 dye removal by a continuous flow electro-coagulation/flotation reactor. *Separation and Purification Technology*, **108** 215-22.
[59] Alizadeh M, Ghahramani E, Zarrabi M and Hashemi S 2015. Efficient De-colorization of Methylene Blue by Electro-coagulation Method: Comparison of Iron and Aluminum Electrode. *Iranian Journal of Chemistry and Chemical Engineering*, **34** 39-47.
[60] Naje A S, Chelliapan S, Zakaria Z and Abbas S A 2016. Electrocoagulation using a rotated anode: A novel reactor design for textile wastewater treatment. *J Environ Manage*, **176** 34-44.
[61] Signorelli S C M, Costa J M and de Almeida Neto A F 2021. Electrocoagulation-flotation for orange II dye removal: kinetics, costs, and process variables effects. *Journal of Environmental Chemical Engineering*, 106157.
[62] Jawad S F, Saddam N S, Adaami Q J, Kareem M M, Abdulredha M, Mubarak H A, Kot P, Gkantou M and AlKhayyat A 2021. Dye removal from textile wastewater using solar-powered electro-coagulation reactor. *IOP Conference Series: Materials Science and Engineering,* 012016.

Effects of organic matter on the performance of water and wastewater treatment: Electrocoagulation a case study

K. Hashim[1,*], Wafaa Ismail Saad[2], K. Safaa[3], Abduljaleel Al-Janabi[3]

[1] College of Environment, Al-Qasim Green University, Babylon, Iraq
[2] College of Education, University of Mustansiriyah, Iraq
[3] Faculty of Engineering, University of Babylon, Iraq

Corresponding author's e-mail: kadhim.hashim@environ.uoqasim.edu.iq

Abstract. One of the commonly available pollutants in freshwater sources is organic matter (OM) because it is abundantly found in nature and the wastewater, where it is believed all forms of freshwater, soils and wastewaters contain a certain concentration of OM. Although the majority of the OM, especially the natural forms, has no serious impacts on the human health or aquatic environment, its high concentrations or its industrial forms can cause health problems along with severe impacts on the treatability of water or wastewater because it chelates the inorganic matters, reacts with other pollutants or covers them, which makes their removal is difficult. Therefore, a vast number of methods were applied to remove OM from water, such as electrocoagulation. Despite the proven efficiency of the EC in the removal of a vast number of pollutants from solutions, its performance is highly influenced by the content of OM in the solution because the OM covers the electrodes decreasing the dissolution rate, and OM also reacts with inorganic matter such as metals making it difficult to remove them. Some solutions were developed to solve these problems, such as adding chloride that inhibits the formation of the inert layers on electrodes, also adding Mg could improve the removal efficiency by maximising the size of the flocs. Thus, this paper reviews the effects of the OM on the performance of EC and lists a number of recent applications of the EC in OM removal.

1. Introduction

1.1. Organic matter chemistry and existence in water

Organic matter (OM) is chemically defined as a group of carbon-containing compounds; in other words, it is a chemical compound/complex that contains carbon atoms in its structure, which can be found in water, soil, and even in the air. OM can be decomposed under the effects of temperature or by living organisms [1, 2]. Therefore, OM is a main component of the aquatic environments, where it could be found in all forms of freshwater, groundwater, reservoirs, domestic sewage, industrial sewage, and agricultural sewages [3-5]

The existence of OM in freshwater results from the interactions between the hydrological cycle and the geosphere and biosphere [6]. Chemically, the structures of the OM generally have carbon atom(s) besides other elements such as H, N and O, depending on the origins of the OM. The concentration and chemical compositions of the OM in water bodies vary pointedly depending on the origin of water or wastewaters [7-9].

Generally, the OM is classified into two types according to their source, synthetic and natural OM. The natural OM [6]. Synthetic OM is usually more complex and harmful than natural OM. For example, phenol, pesticide, and organic acids are synthetic OM and could be found in industrial and agricultural sewages [10-12] while the natural OM is derived from the residuals of plants and animals and humic substances and enter the aquatic environment from the surrounding environments, such as the banks of the river or the bed soil [13, 14] Additionally, the concentration and composition of the OM may seasonally differ in the same place within the aquatic environments [15, 16]. This difference in concentration and type of OM applies noteworthy effects on selections, designs and performances of treatment means [17-19].

Furthermore, the concentrations of the OM in water and sewages are increasing day by day because of the increase in world population. For example, the global population has increased to about 7 billion people during the last century [20, 21]; this increase was accompanied by a vast increase in urbanisations, industry and agriculture [22-24]. Consequently, the usage and disposal of OM have increased many folds [22-26], where the need for new houses, infrastructures, food and freshwater supplies contributed to the increase in production, use and disposal of the OM. An example of this rapid increase in the use of OM is the textile industry that consumes huge amounts of OM in both fabrics and dyes, where the number of the synthetic dyes that are currently used in the textile industry is more than 100×10^3 types and the global production of these dyes is more than 100×10^3 tons/year. Taking into account these vast amounts of dyes will be mixed with bigger volumes of water (in the range of 100-200 L per kg of fabrics), and wasted water will be discharged back to the rivers and lakes. Another example of pollution with OM is the refineries, where the estimated daily global sewage discharge from the refining industry is about 40×10^6 litres [4, 27]. Another reason for the pollution of water with organic matter is the ladfilling of manucipal solid wastes [28], this source of pollution is incraeing due to the expansion of cities that dispose huge voluems of solid wastes everyday [29-31]. Climate change is also causes huge increase in the OM pollution because it either decreases the flow of rivers [32, 33], or cuases floods in other places [34-36], which results in huge pollution and water shortage consequently [37-39].

Thus, OM pollution represents a challenge to water and wastewater treatment utilities [40, 41] Several treatment approaches were experienced to remove the OM from sewages before effluent them to the surface water, such as the aerobic and anaerobic degradations [42, 43], adsorptions [27, 44], reverse osmosis and other filtration methods [45, 46], and electrocoagulation (EC) methods [46-48].

The current study focuses on the EC method because it cheap, safe, and efficient method that can be utilised even in low-income countries. At the same time, it is highly influenced by the OM, which makes it a good example of the topic of this study.

1.2. Measurement of OM concentration

The concentration of OM in water is measured by two methods, namely biochemical oxygen demand (BOD) and chemical oxygen demand (COD) [28].

BOD is defined as the needed amount of O_2, usually expressed in mg/L or ppm units, for the living organisms to oxidise OM in solution. This test is usually done at room temperature for a specific period (usually 5 days). While, the COD is an equivalent measure of the required O_2 to decompose the OM in a sample of the solution, and it is done using strong chemical oxidants. COD test has many advantages; for instance, it is simple and takes a much shorter time (about 2 hrs) in comparison to the BOD test that normally requires 5 days [28].

1.3. Influence of OM on quality and treatability of water

The attendance of OM in freshwater or sewages has important negative influences on organoleptic properties and the treatability of that solution [49, 50]. For example, OM endorses the growth of the microbial, consumption coagulants, decreases the effects of water disinfectants, and it reacts with most of the chemical pollutants in nature, forming pollutants that are less disposed to coagulate, and it produces unwanted colour, taste, and odour [47, 51].

In addition, a wide body of studies has shown that OM causes grave health problems, such as cancer, due to its ability to react with water disinfectants, like Cl, forming toxic disinfection by-products such as trihalomethanes and haloacetic [52, 53].

Because of these thoughtful influences, the health and environmental agencies strictly regulated OM and its by-products in drinking water. For example, the European Union countries limited BOD and THM in drinking water to 6 mg/L and 0.1 mg/L, respectively, while the USEPA limits THM and HAA in drinking water to 0.08 and 0.06 mg/L [28].

Many treatment methods, therefore, were used to remove the OM and its by-products from water or wastewater.

1.4. OM treatment methods

To meet the limitations for OM and its by-products in drinking water, many treatment methods have been used to remove OM from drinking water, such as [28, 51]:

1- Chemical coagulation
2- Adsorption
3- membrane filtration
4- Biological digestion
5- EC

Unfortunately, the previous studies showed there is no single method alone that can completely remove the OM and its by-products because of the high variation in their concentrations and chemical compositions [28].

For instance, OM fouls of the membranes that harmfully effects the efficiency of the membrane filtration process [8]. It negatively affects the coagulation process because OM reacts with most of the pollutants to form complexes less disposed to coagulate, and it upsurges the consumption of coagulants. Finally, OM negatively affects the adsorption process due to blocking the adsorption sites on the surfaces of adsorbents, which meaningfully decreases the removal efficiency [8].

In the EC method, the OM decreases the dissolution rate of electrodes, reacts with most of the pollutants to form complexes less disposed to coagulate and increase the energy consumption. However, the EC method is still one of the most affordable treatment methods; therefore, this method will be the focus of this paper.

1.5. EC method

The common defection of the EC is the in situ production of coagulants through the dissolution of metallic electrodes under the effects of direct electric current, without the need for chemical coagulants or agents [46]. The EC method is consisting of sets of electrodes, anodes and cathodes, connected to a direct electric current source. The DC current motivates the anode to produce positively charged ions and the cathode to produce hydrogen gas. The ions travel towards the negative electrode (the cathode); during this travel, the negative ions react with the pollutants in the solution (they usually have a positive charge), forming a floc, the latter grow in size until it reaches a heavyweight that can not be carried by the solution anymore [27]. As a result, the heavy flocs will be removed from the solution via precipitation and collected from the bottom of the unit as sludge. Another removal scenario happens when the applied current is high, which motivates the cathode to produce more hydrogen gas in bubbles. Because the produced bubbles are lighter than the solution, they will move upward to leave the solution; during this travel, flocs will be attached to these bubbles and accumulated on the surface of the solution as foam, which will be scammed later using a metallic plate [4].

When the direct current is applied, the electrocoagulation reaction started, and these reactions depend on many factors, such as the type of electrodes and pH of the solution. Figure 1 shows a general EC unit.

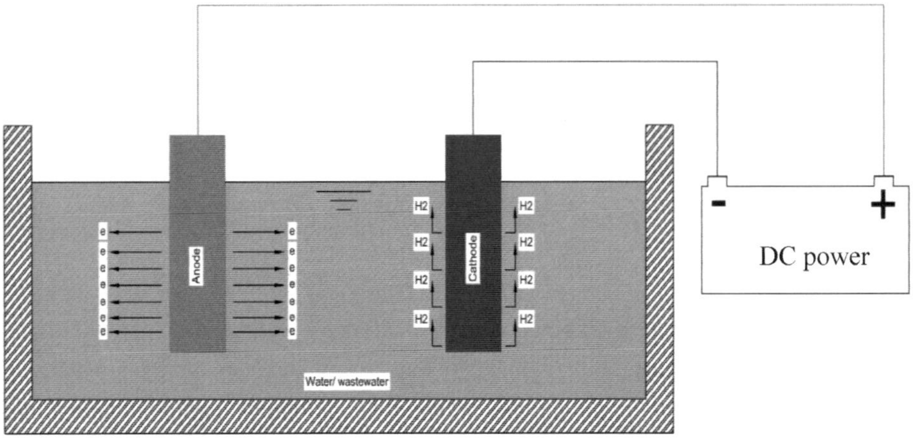

Figure 1: General arrangement of the EC method.

1.6. Effects of OM on the EC method

In spite of the acknowledged advantages of EC technology to treat a wide spectrum of pollutants from water and wastewater, its efficiency is considerably influenced by some parameters such as the chemistry of the solution being treated [51]. Due to the high reactability of OM with most of the chemicals found in the water forming different complexes, and thereby changing the chemistry of water, OM significantly influences the performance of the EC method. For instance, Fan, Tseng, Li and Hou [54] used carbon electrodes to treat 0.2 mg/L arsenic-containing water samples. The latter reported that the presence of 20 mg/L of OM was enough to reduce arsenic removal efficiency by about 45%.

Generally, the impacts of the OM on the EC method can be summarised as follows [27, 28, 41]:

- ✓ OM chelates some of the target inorganic pollutants, forming a complex that is difficult to remove, which in turn reduces the removal efficiency.
- ✓ In iron-based electrodes, OM tends to react rapidly with the freshly generated ferrous coagulation ions, a process that disables flocs formation.
- ✓ OM coats the inorganic pollutants that inhibit their coagulation.
- ✓ It promotes the growth of the passive oxide film on the anode that decreases the number of cations at the anode and consequently minimises the rate of floc formation and the adsorption of pollutants.
- ✓ OM reacts with chlorides, which is important to enhance water conductivity and to inhibit the formation of the anodic layer, forming toxic by-products.
- ✓ The presence of OM reduces the size of the formed flocs because of the more negative zeta potential, which in turn negatively influences the coagulation process.
- ✓ OM competes for active adsorption sites on flocs, decreasing the removal of the targeted pollutants.
- ✓ Fractionation of high molecular weight organic compounds during EC produces hydrophilic and low molecular weight compounds, leading to a poor coagulation process.
- ✓ OM increases the energy consumption of the EC units due to promoting the growth of an impermeable oxide layer on the anode, which in turn resists the electric current.

A good example of the OM impacts on the EC method is the heavy metals removal from drinking water in the presence of OM. For instance, using EC technology, Kobya, Akyol, Demirbas and Oncel [55] reduced arsenic concentration in drinking water from 150 to 6.3 µg/L (96%) in less than 6 min retention time at a current density of 0.25 mA/cm^2, and A/V (electrodes area/volume of water) of 0.24 cm^{-1}. However, in the presence of 9.31 ± 0.51 mg/L of organic matter, Mohora, Roncevic, Dalmacija, Agbaba, Watson, Karlovic and Dalmacija [56] needed, with the same A/V, 90 min of electrolysis at a current density of 5.78 mA/cm^2 to reduce the arsenic concentration from only 45 to 6.2 µg/L (85%). Furthermore, Mohora, Roncevic, Dalmacija, Agbaba, Watson, Karlovic and Dalmacija [56] added an additional 60 mg/L of Cl^- to an EC unit to break down the anodic oxide layer promoted by the presence of the OM. This example reveals that the presence of OM increased the required retention time and the consumed energy by 15 and 23-fold, respectively, to remove 1/3 of the arsenic concentration that was removed in the absence of OM. Moreover, Mohora, Roncevic, Dalmacija, Agbaba, Watson, Karlovic and Dalmacija [56] used high chloride concentration to enhance the performance of the EC unit, which gives a high probability of forming toxic by-products.

Due to the mentioned significant impacts of the OM on water treatability and performance of treatment methods, this study focuses on developing a new, cost-effective, and environmentally friendly hybrid method to remove heavy metals from water in the presence of OM. This method is a combination of microwave and EC technologies (microwave-assisted- EC). Therefore, for a better understanding of MW effects on the removal of water pollutants, it would be necessary to allocate the following chapter to explain the fundamentals, theory, and applications of MW in water and wastewater treatment.

1.7. Uses of the EC method in the removal of OM from solutions

Table 1 lists a number of the recent studies that dealt with the use of the EC in the removal of OM from solutions.

Table 1: Studies used the EC method to remove OM from solutions.

Type of water or wastewater	Material of electrode	Type of reactor	Optimum operating conditions	Removal efficiency	Authors
Olive mill wastewater	Aluminium	Batch	Time = 25 min, current density = 75 mA/cm^2, initial pH = 4-6, initial COD = 75.1 g/L, power consumption = Not given, gap between electrodes = 20 mm.	76% of COD	[57]
wastewater of potato chips plant	Aluminium and iron	Batch	The best removal by Al electrodes, time = 40 min, current density = 20 mA/cm^2, initial pH = 4, initial COD = 2.2-2.8 g/L, power consumption = 4 kWh/m^3, gap between electrodes = 11 mm.	60% of COD	[58]
synthetic waters	Iron cast	Batch	Time = 70 min, applied voltage = 50V, initial pH = 9, initial concentration of humic acid = 200 mg/L, power	92.69% of COD	[59]

			consumption ≈ 10 kWh/m^3, gap between electrodes = 50 mm.		
Wastewater of agro-industry	Mild steel and aluminium	Batch	Time = 60 min, current density = 5 mA/cm^2, final pH =7.6, power consumption = up to 51.6 kWh/m^3 (according to the type of wastewater), gap between electrodes = 15 mm.	About 85% of COD using bipolar mild steel electrodes	[60]
Wastewater of paper mill	Aluminium and iron	Batch	Time = 7.5 min, current density = 4.8 mA/cm^2, initial pH =7.6, initial concentration of phenol and lignin is 0.535 and 13.514 mg/L respectively, power consumption = 95.52 Kw/m^3, gap between electrodes = 20 mm.	Al electrodes were 80% of lignin and 98% of phenol. Fe electrodes 92% of lignin and 93% of phenol.	[61]
Wastewater of vegetable oil refinery	Aluminium	Batch	Time = 90 min, current density = 35 mA/cm^2, initial pH =7, power consumption = 42.6 kWh/kg of COD, gap between electrodes = 8 mm.	98.9% of COD	[62]
Hospital wastewater	Fe-Fe Al-Al Al-Fe	Batch	Time = 60 min, applied voltage = 30V, initial pH =3, power consumption = 30.6 kWh/m^3, gap between electrodes = 20 mm.	87.1% of COD	[63]
Wastewater of petroleum refinery	Iron as anode and aluminium as a cathode	Batch	Time = 30 min, current density = 12.2 mA/cm^2, initial pH =9, initial COD = 72,450 mg/L, power consumption = not given, gap between electrodes = 10 mm.	80% of COD	[64]
Wastewater of distillery industries	Iron and aluminium	Batch	Iron electrodes gave the highest removal efficiency at time = 150 min, applied voltage = 25V, initial pH =3, initial COD = 110-190 g/L,	85.7% of COD	[65]

			power consumption = not given, gap between electrodes = not given.		
Seawater	aluminium	Batch	Time = 40 min, current density = 22.4 mA/cm^2, initial pH =3, initial COD = 1.34 mg/L, power consumption = not given, gap between electrodes = 10 mm.	70.8% of COD	[66]
Synthetic oil refinery wastewater	Iron as anode and aluminium as a cathode	Batch	Time = 40 min, applied voltage = 10.5V, initial pH =7, initial diesel concentration = 3.5 g/L, power consumption = 6.47 kWh/m^3, gap between electrodes = 20 mm.	98% of COD	[67]

2. Conclusion

The final summary of the review is, firstly, the EC method is eco-friendly as it does depend on metallic plates to provide the coagulants instead of the additions of external chemicals; this means no secondary pollutants will be produced. Secondly, the EC is easy to be operated without the need for skilled staff.

However, the chemical composition of the solution affects the efficiency of the EC method, especially the OM, because the OM can chelate some of the target inorganic pollutants forming a complex that is difficult to remove; OM tends to react rapidly with the freshly generated ferrous coagulation ions, a process that disables flocs formation. OM coats the inorganic pollutants that inhibit their coagulation. OM promotes the growth of the passive oxide film on the anode that decreases the number of cations at the anode and consequently minimises the rate of floc formation and the adsorption of pollutants.

These problems could be solved by adding some chemicals or by supplying the EC units with pre-treatment units to minimise the concentration of OM.

References

[1] Tavali I E. Evaluation of the effects of organic amendments on soil microbial dynamics in terms of sustainable soil fertility. *Academic Research and Reviews in Agriculture, Forestry and Aquaculture Sciences*, 43.

[2] Hashim K S, Khaddar R A, Jasim N, Shaw A, Phipps D, Kot P, Pedrola M O, Alattabi A W, Abdulredha M and Alawsh R 2019. Electrocoagulation as a green technology for phosphate removal from River water. *Separation and Purification Technology*, **210** 135-44.

[3] Hashim K S, Shaw A, Al Khaddar R, Ortoneda Pedrola M and Phipps D 2017. Defluoridation of drinking water using a new flow column-electrocoagulation reactor (FCER) - Experimental, statistical, and economic approach. *Journal of Environmental Management*, **197** 80-8.

[4] Abdulhadi B A, Kot P, Hashim K S, Shaw A and Khaddar R A 2019. Influence of current density and electrodes spacing on reactive red 120 dye removal from dyed water using electrocoagulation/electroflotation (EC/EF) process. *First International Conference on Civil and Environmental Engineering Technologies (ICCEET),* University of Kufa, Iraq 12-22.

[5] Omran I I, Al-Saati N H, Hashim K S, Al-Saati Z N, Patryk K, Khaddar R A, Al-Jumeily D, Shaw A, Ruddock F and Aljefery M 2019. Assessment of heavy metal pollution in the Great Al-Mussaib irrigation channel. *Desalination and Water Treatment*, **168** 165-74.

[6] Hashim K S, AlKhaddar R, Shaw A, Kot P, Al-Jumeily D, Alwash R and Aljefery M H 2020 *Electrocoagulation as an eco-friendly River water treatment method. In Advances in Water Resources Engineering and Management*, Berline: Springer.

[7] Mohammed A-H, Hussein A H, Yeboah D, Al Khaddar R, Abdulhadi B, Shubbar A A and Hashim K S 2020. Electrochemical removal of nitrate from wastewater. *IOP Conference Series: Materials Science and Engineering,* 012037.

[8] Abdulhadi B, Kot P, Hashim K, Shaw A, Muradov M and Al-Khaddar R 2021. Continuous-flow electrocoagulation (EC) process for iron removal from water: Experimental, statistical and economic study. *Science of The Total Environment*, **760** 1-16.

[9] Emamjomeh M M, Mousazadeh M, Mokhtari N, Jamali H A, Makkiabadi M, Naghdali Z, Hashim K S and Ghanbari R 2020. Simultaneous removal of phenol and linear alkylbenzene sulfonate from automotive service station wastewater: Optimization of coupled electrochemical and physical processes. *Separation Science and Technology*, **55** 3184-94.

[10] Hashim K, Kot P, Zubaid S, Alwash R, Al Khaddar R, Shaw A, Al-Jumeily D and Aljefery M 2020. Energy efficient electrocoagulation using baffle-plates electrodes for efficient Escherichia Coli removal from Wastewater. *Journal of Water Process Engineering*, **33** 101079-86.

[11] Hashim K S, Ali S S M, AlRifaie J K, Kot P, Shaw A, Al Khaddar R, Idowu I and Gkantou M 2020. Escherichia coli inactivation using a hybrid ultrasonic–electrocoagulation reactor. *Chemosphere*, **247** 125868-75.

[12] Hashim K S, Ewadh H M, Muhsin A A, Zubaidi S L, Kot P, Muradov M, Aljefery M and Al-Khaddar R 2020. Phosphate removal from water using bottom ash: Adsorption performance, coexisting anions and modelling studies. *Water Science and Technology*, **83** 1-17.

[13] Alenezi A, Hashim K, Amoako-Attah J, Gkantou M, Muradov M, Kot P and Abdulhadi B 2020. Zeolite-assisted electrocoagulation for remediation of phosphate from calcium-phosphate solution. *IOP Conference Series: Materials Science and Engineering,* 012031.

[14] Zanki A K, Mohammad F H, Hashim K S, Muradov M, Kot P, Kareem M M and Abdulhadi B 2020. Removal of organic matter from water using ultrasonic-assisted electrocoagulation method. *IOP Conference Series: Materials Science and Engineering,* 012033.

[15] Alenazi M, Hashim K S, Hassan A A, Muradov M, Kot P and Abdulhadi B 2020. Turbidity removal using natural coagulants derived from the seeds of strychnos potatorum: statistical and experimental approach. *IOP Conference Series: Materials Science and Engineering,* 012064.

[16] Alhendal M, Nasir M J, Hashim K S, Amoako-Attah J, Al-Faluji D, Muradov M, Kot P and Abdulhadi B 2020. Cost-effective hybrid filter for remediation of water from fluoride. *IOP Conference Series: Materials Science and Engineering,* 012038.

[17] Abdulla G, Kareem M M, Hashim K S, Muradov M, Kot P, Mubarak H A, Abdellatif M and Abdulhadi B 2020. Removal of iron from wastewater using a hybrid filter. *IOP Conference Series: Materials Science and Engineering,* 012035.

[18] Abdulraheem F S, Al-Khafaji Z S, Hashim K S, Muradov M, Kot P and Shubbar A A 2020. Natural filtration unit for removal of heavy metals from water. *IOP Conference Series: Materials Science and Engineering,* 012034.

[19] Hashim K, Hussein A, Zubaidi S, Kot P, Kraidi L, Alkhaddar R, Shaw A and Alwash R 2019. Effect of initial pH value on the removal of reactive black dye from water by electrocoagulation (EC) method. *2nd International Scientific Conference,* Al-Qadisiyah University, Iraq 12-22.

[20] Hashim K, Al-Saati N and Hussein A 2018. An investigation into the level of heavy metals leaching from canal-dreged sediment: a case study metals leaching from dreged sediment. *1st Int. Conference on Materials Engineering & Science,* Istanbul Aydın University, Turkey 12-22.

[21] Hashim K S, Idowu I A, Jasim N, Al Khaddar R, Shaw A, Phipps D, Kot P, Pedrola M O, Alattabi A W and Abdulredha M 2018. Removal of phosphate from River water using a new baffle plates electrochemical reactor. *MethodsX*, **5** 1413-8.
[22] Al-Jumeily D, Hashim K, Alkaddar R and Lunn J 2019. Sustainable and Environmental Friendly Ancient Reed Houses (Inspired by the Past to Motivate the Future). *11th International Conference on Developments in eSystems Engineering (DeSE),* Cambridge, UK 214-9.
[23] Farhan S L, Hashim I A J and Naji A A 2019. The Sustainable House: Comparative Analysis of Houses in Al Kut Neighborhoods-Iraq. *2019 12th International Conference on Developments in eSystems Engineering (DeSE),* 1031-6.
[24] Farhan S L, Antón D, Akef V S, Zubaidi S L and Hashim K S 2021. Factors influencing the transformation of Iraqi holy cities: the case of Al-Najaf. *Scientific Review Engineering and Environmental Sciences*, **30** 365-75.
[25] Al-Saati N H, Omran I I, Salman A A, Al-Saati Z and Hashim K S 2021. Statistical modeling of monthly streamflow using time series and artificial neural network models: Hindiya Barrage as a case study. *Water Practice and Technology*, **16** 681-91.
[26] Al-Sareji O J, Grmasha R A, Salman J M, Idowu I and Hashim K S 2021. Street dust contamination by heavy metals in Babylon governorate, Iraq. *Journal of Engineering Science and Technology*, **16** 3528 - 46.
[27] Hashim K, Al-Saati N, Alquzweeni S, Zubaidi S, Kot P, Kraidi L, Hussein A, Alkhaddar R, Shaw A and Alwash R 2019. Decolourization of dye solutions by electrocoagulation: an investigation of the effect of operational parameters. *First International Conference on Civil and Environmental Engineering Technologies (ICCEET),* University of Kufa, Iraq 25-32.
[28] Hashim K S 2017 The innovative use of electrocoagulation-microwave techniques for the removal of pollutants from water. PhD thesis, Liverpool, United Kingdom: Liverpool John Moores University, UK, p 229
[29] Abdulredha M, Rafid A, Jordan D and Hashim K 2017. The development of a waste management system in Kerbala during major pilgrimage events: determination of solid waste composition. *Procedia Engineering*, **196** 779-84.
[30] Abdulredha M, Al Khaddar R, Jordan D, Kot P, Abdulridha A and Hashim K 2018. Estimating solid waste generation by hospitality industry during major festivals: A quantification model based on multiple regression. *Waste Management*, **77** 388-400.
[31] Idowu I A, Atherton W, Hashim K, Kot P, Alkhaddar R, Alo B I and Shaw A 2019. An analyses of the status of landfill classification systems in developing countries: Sub Saharan Africa landfill experiences. *Waste Management*, **87** 761-71.
[32] Zubaidi S, Al-Bugharbee H, Muhsen Y, Hashim K, Alkhaddar R, Al-Jumeily D and Aljaaf A 2019. The Prediction of Municipal Water Demand in Iraq: A Case Study of Baghdad Governorate. *12th International Conference on Developments in eSystems Engineering (DeSE),* Kazan, Russia 274-7.
[33] Zubaidi S L, Kot P, Hashim K, Alkhaddar R, Abdellatif M and Muhsin Y R 2019. Using LARS–WG model for prediction of temperature in Columbia City, USA. *IOP Conference Series: Materials Science and Engineering,* 012026.
[34] Salah Z, Abdulkareem I, Hashim K, Al-Bugharbee H, Ridha H, Al-Qaim F, Muradov M and Kot P 2020. Hybridised Artificial Neural Network model with Slime Mould Algorithm: A novel methodology for prediction urban stochastic water demand. *Water*, **12** 1-18.
[35] Salah Z, Hashim K, Ethaib S, Al-Bdairi N, Al-Bugharbee H and Gharghan S 2020. A novel methodology to predict monthly municipal water demand based on weather variables scenario. *Journal of King Saud University-Engineering Sciences*, **32** 1-18.
[36] Salah Z, Ortega-Martorell S, Kot P, Alkhaddar R M, Abdellatif M, Gharghan S, Ahmed M and Hashim K 2020. A Method for Predicting Long-Term Municipal Water Demands Under Climate Change. *Water Resources Management*, **34** 1265-79.

[37] Zubaidi S, Al-Bugharbee H, Muhsin Y R, Hashim K and Alkhaddar R 2020. Forecasting of monthly stochastic signal of urban water demand: Baghdad as a case study. *IOP Conference Series: Materials Science and Engineering,* 012018.

[38] Zubaidi S, Ortega-Martorell S, Al-Bugharbee H, Olier I, Hashim K S, Gharghan S K, Kot P and Al-Khaddar R 2020. Urban Water Demand Prediction for a City that Suffers from Climate Change and Population Growth: Gauteng Province case study. *Water*, **12** 1-18.

[39] Zubaidi Salah L, Al-Bugharbee H, Ortega Martorell S, Gharghan S, Olier I, Hashim K, Al-Bdairi N and Kot P 2020. A Novel Methodology for Prediction Urban Water Demand by Wavelet Denoising and Adaptive Neuro-Fuzzy Inference System Approach. *Water*, **12** 1-17.

[40] Hashim K S, Shaw A, Al Khaddar R, Pedrola M O and Phipps D 2017. Iron removal, energy consumption and operating cost of electrocoagulation of drinking water using a new flow column reactor. *Journal of Environmental Management*, **189** 98-108.

[41] Al-Marri S, AlQuzweeni S S, Hashim K S, AlKhaddar R, Kot P, AlKizwini R S, Zubaidi S L and Al-Khafaji Z S 2020. Ultrasonic-Electrocoagulation method for nitrate removal from water. *IOP Conference Series: Materials Science and Engineering,* 012073.

[42] Alattabi A W, Harris C, Alkhaddar R, Alzeyadi A and Hashim K 2017. Treatment of Residential Complexes' Wastewater using Environmentally Friendly Technology. *Procedia Engineering*, **196** 792-9.

[43] Alattabi A W, Harris C B, Alkhaddar R M, Hashim K S, Ortoneda-Pedrola M and Phipps D 2017. Improving sludge settleability by introducing an innovative, two-stage settling sequencing batch reactor. *Journal of Water Process Engineering*, **20** 207-16.

[44] Al-Saati N, Hussein T K, Abbas M, Hashim K, Al-Saati Z, Kot P, Sadique M, Aljefery M and Carnacina I 2019. Statistical modelling of turbidity removal applied to non-toxic natural coagulants in water treatment: a case study. *Desalination and Water Treatment*, **150** 406-12.

[45] Alyafei A, AlKizwini R, Hashim K, Yeboah D, Gkantou M, Al-Faluji D and Zubaidi S L 2020. Treatment of effluents of construction industry using a combined filtration-electrocoagulation method. *IOP Conference Series: Materials Science and Engineering,* 012032.

[46] Emamjomeh M, Kakavand S, Jamali H, Alizadeh S, Safdari M, Mousavi S, Hashim K and Mousazade M 2020. The treatment of printing and packaging wastewater by electrocoagulation–flotation: the simultaneous efficacy of critical parameters and economics. *Desalination and water treatment*, **205** 161-74.

[47] Aqeel K, Mubarak H A, Amoako-Attah J, Abdul-Rahaim L A, Al Khaddar R, Abdellatif M, Al-Janabi A and Hashim K S 2020. Electrochemical removal of brilliant green dye from wastewater. *IOP Conference Series: Materials Science and Engineering,* 012036.

[48] Grmasha R, Al-sareji O, Salman J, Hashim K and Jasim I 2020. Polycyclic Aromatic Hydrocarbons (PAHs) in Urban Street Dust Within Three Land-Uses of Babylon Governorate, Iraq: Distribution, Sources, and Health Risk Assessment. *Journal of King Saud University - Engineering Sciences*, **33** 1-18.

[49] Hashim K, Shaw A and Phipps D, 2019. *Treatment reactor and method of treating a liquid.* Patent, WIPO, PCT/GB2019/052493, LJM University, United Kingdom.

[50] Alnaimi H, Idan I J, Al-Janabi A, Hashim K, Gkantou M, Zubaidi S L, Kot P and Muradov M 2020. Ultrasonic-electrochemical treatment for effluents of concrete plants. **888** 1-9.

[51] Hashim K S, Shaw A, AlKhaddar R, Kot P and Al-Shamma'a A 2021. Water purification from metal ions in the presence of organic matter using electromagnetic radiation-assisted treatment. *Journal of Cleaner Production*, **280** 1-17.

[52] Hashim K, Shaw A, Pedrola M and Phipps D 2017. Energy efficient electrocoagulation using a new flow column reactor to remove nitrate from drinking water - Experimental, statistical, and economic approach. *Journal of Environmental Management*, **196** 224-33.

[53] Omran I I, Al-Saati N H, Al-Saati H H, Hashim K S and Al-Saati Z N 2021. Sustainability assessment of wastewater treatment techniques in urban areas of Iraq using multi-criteria decision analysis (MCDA). *Water Practice and Technology*, **16** 648-60.

[54] Fan C S, Tseng S C, Li K C and Hou C H 2016. Electro-removal of arsenic(III) and arsenic(V) from aqueous solutions by capacitive deionization. *J Hazard Mater*, **312** 208-15.
[55] Kobya M, Akyol A, Demirbas E and Oncel M S 2014. Removal of arsenic from drinking water by batch and continuous electrocoagulation processes using hybrid Al-Fe plate electrodes. *Environmental Progress & Sustainable Energy*, **33** 131-40.
[56] Mohora E, Roncevic S, Dalmacija B, Agbaba J, Watson M, Karlovic E and Dalmacija M 2012. Removal of natural organic matter and arsenic from water by electrocoagulation/flotation continuous flow reactor. *J Hazard Mater*, **235-236** 257-64.
[57] Adhoum N and Monser L 2004. Decolourization and removal of phenolic compounds from olive mill wastewater by electrocoagulation. *Chemical Engineering and Processing: Process Intensification*, **43** 1281-7.
[58] Kobya M, Hiz H, Senturk E, Aydiner C and Demirbas E 2006. Treatment of potato chips manufacturing wastewater by electrocoagulation. *Desalination*, **190** 201-11.
[59] Yildiz Y S, Koparal A S, Irdemez S and Keskinler B 2007. Electrocoagulation of synthetically prepared waters containing high concentration of NOM using iron cast electrodes. *J Hazard Mater*, **139** 373-80.
[60] Drogui P, Asselin M, Brar S, Benmoussa H and Blais J 2008. Electrochemical removal of pollutants from agro-industry wastewaters. *Separation and Purification Technology*, **61** 301.
[61] Ugurlu M, Gurses A, Dogar C and Yalcin M 2008. The removal of lignin and phenol from paper mill effluents by electrocoagulation. *J Environ Manage*, **87** 420-8.
[62] Un U, Koparal A S and Bakir Ogutveren U 2009. Electrocoagulation of vegetable oil refinery wastewater using aluminum electrodes. *J Environ Manage*, **90** 428-33.
[63] Dehghani M, Seresht S S and Taghizadeh M 2012. Optimization of organic compounds removal from Wastewater by Electrocoagulation. *Hormozgan Medical Journal*, **19**
[64] Ben I, Halleb A, Adhoum N and Monser L 2013. Treatment of petroleum refinery sulfidic spent caustic wastes by electrocoagulation. *Separation and Purification Technology*, **107** 150-7.
[65] Wagh M, Nemade P and Dhasal S 2015. Colour and COD removal of Distillery spent wash by using Electro coagulation. *International Journal of Engineering Research and General Science*, **3** 1159-73.
[66] Hakizimana J N, Gourich B, Vial C, Drogui P, Oumani A, Naja J and Hilali L 2016. Assessment of hardness, microorganism and organic matter removal from seawater by electrocoagulation as a pretreatment of desalination by reverse osmosis. *Desalination*,
[67] Safari S, Azadi Aghdam M and Kariminia H R 2016. Electrocoagulation for COD and diesel removal from oily wastewater. *International Journal of Environmental Science and Technology*, **13** 231-42.

Reduction of phosphates in sewage using kiln ash

D Mahmood[1], A. Adnan[1,*]
[1] Department of Environment, Ministry of Oil, Iraq

* Corresponding author; Email: adnanhilla2021@gmail.com

Abstract. Phosphate is among few chemicals that abundantly occur in both soil and water environments at relatively high concentrations, and it is responsible for the promoted growth of algae in a phenomenon called eutrophication. Hence, many studies have been conducted to assess the ability of various treatments to remove phosphates from sewage. Recently, it has been discovered that phosphorus recovery may be accomplished by filtering the water. The cost of the filtering materials, on the other hand, is prohibitively expensive. As a result, current research has concentrated on utilising low-cost ones to minimise the expense of filtering. Steel production residues, such as kiln bottom ashes, are being utilised in this study to recover phosphates from contaminated wastewater. A variety of operational settings have been investigated in order to obtain the highest possible extraction efficiency at the lowest feasible cost. The results of this study showed 0.53 g/L of bottom ash needed not more than 40 minutes to reduce the phosphate concentration from 5 to 4.51 mg/L. Additionally, the results were modelled with an R2 value of more than 0.9, confirming the agreement between the theoretical and experimental results.

1. Introduction

The previous studies showed elevated concentrations of phosphate in nature (water and soil); however, urbanisation and industrial expansion drive huge global demand for drinkable water that exhausts the available freshwater rapidly, and in return, increasing the volumes of discharged wastewater that contains high concentrations of phosphate [1, 2]. In addition, the urban spread and industrial expansion are badly reflected in the availability of freshwater [3-5], global warming [6, 7], and precipitations [8, 9]. The growing population increases the rate of sewage released into waterways, which has severe effects on the health of consumers [10-12]. Furthermore, the increase in the population increased the dumping of solid wastes in landfills that pollute groundwater and then surface water [13-16]. Toxins must not be dumped directly into waterways, and all effluents, whether household or manufacturing, must be treated adequately to eliminate or limit the negative impacts of the contaminants. One of the types of impact of pollutants is algae flourishing which is due to phosphorous penetration into the water, which is harmful to the environment [17]. It is proved that the majority of sewage water contain phosphorus [18, 19] and nitrate [20, 21] besides other pollutants such as fluoride [22], heavy metals [23, 24], bacteria [25, 26], colouring pollutants [27-29] and organic matter [30-32]. Additionally, industrial and domestic wastewater [33-36] and atmospheric pollutants [37-39] are playing a role in the increasing of such pollutants in water bodies. The effluents of the construction industry [40-43], production plants [44-46] and food industry [33, 47].

The main effect of phosphate on freshwater sources is the depletion of oxygen due to algal blooms induced by phosphorus and nitrate contamination [12]. A large proportion of aquatic creatures and plants might die in a short period of time. As just a result of a substantial rise in organic matter content, the taste, smell, and structure of the water will be changed significantly

[18, 19]. A rise in algal development might have severe effects on health, industries, seafood, and agriculture, and farmers [21, 48]. Many studies revealed that microalgae types can generate poisonous chemicals that cause a range of severe illnesses and, sometimes, fatality. In drinkable water, the allowable concentration of phosphate is specified at 0.1 mg/L, whereas in streams and ponds, it is 0.05 mg/L [18, 19]. As far as the upper limit pollutant concentration in water is concerning, the Environmental Protection Agency has strict standards. It is therefore important to remove phosphates from contaminated water, depending on what has been said previously. When phosphate is removed via traditional methods, it generally requires a long and complicated procedure. As an instance, the normal filtering process [22], with some modifications, provides an effective approach for phosphate removal, according to the authors. However, the usage of expensive screens makes this approach uneconomical [47, 49].

It is possible to remove phosphate from the water bodies by converting it to a solid-state, which would be subsequently removed from the water via flotation or some other means. Thermal processing, lakes, anaerobic digestion, absorption, and other techniques are used to transform phosphorus [19].

The biological treatment methods for phosphate removal depend on bacteria and microalgae species to digest phosphate in solutions. Depends on the kind of microbe, anaerobic or aerobic techniques could be used in these situations. The pH, starting dosage, etc., are all factors that affect the efficacy of the treatments. In regards to employing biological intervention to reduce the existence of phosphate, several disadvantages are discovered [18]. Soluble carbon, fast development of microorganisms, and the pH level of the fluid are all factors to be considered. Phosphate particles can be converted by adding metallic compounds to water polluted with Phosphorus particles. Usually, sewer water encompassing phosphate is treated using chemical techniques. Calcium ion is amongst the most frequently utilised chemical additions for phosphorus removal for its availability, affordability, and convenience of usage. Especially in an acidic medium, aluminium hydroxide is believed to be highly successful in eliminating phosphate from wastewater. As a consequence of the usage of chemical substances, chemical treatment methods frequently lead to toxic sludge [21]. Due to the utilisation use reagents, the technique is much more costly than many other procedures. Physically, phosphate extraction methods do not change the chemical composition of the phosphate, which is resulting in a change. After the removal of phosphate pollution using this approach, it is possible to separate these contaminants (phosphate contaminants) from the removal medium. For phosphate removal, filtering and screening technologies are frequently employed [18, 19].

Additionally, phosphate removal from water could also be achieved using other approaches such as coagulation [10, 28], absorption [11, 50], and straining hybrid methods [20, 30]. To remove phosphates from water, several permeable mediums were employed. The general rule, the expense of filtering is determined based on the cost of the adsorbent materials employed. The phosphate removal capability of membranes has been demonstrated in previous studies. There are several methods to extract phosphates from water at low levels, but reverse osmosis is among the most popular. Polluted water is forced through a barrier with really limited porosity under high pressure. Assist in removing pollutants in water by means of high-pressured water. Membrane systems are rarely utilised due to their expensive running costs, difficulties in discarding them, and the requirement for specific equipment and expertise. Recently, studies have centred on the use of waste products or natural resources as economical and useful replacements for traditional materials [40, 51]. Fly ash is a standard manufacturing byproduct that would be used to remove phosphate from wastewater and also is widely used in the construction industry [52-54], and in concrete industry [55, 56]. Water with an acidity level of 7 can be effectively treated with this substance. Due to the above, our research focused primarily on the possibility of using furnaces bottom ashes as an absorbent to extract the phosphates from polluted effluent.

2. Methods

Depending on the procedures of the previous studies that used bottom ash to remove phosphate from the wastewater, the most important parameters in the adsorption process is the grain size, permeability, porous structure, and relative density of the bottom ashes is essential. The particle size is highly essential because it is directly related to the adsorption surface. When sieve review is performed, sieve sizes of 4, 6, 10, 16, 30 and 50 are utilised, and the quantity of mass kept on each sieve is calculated by analysing the distribution of particle size of the waste. This formula is used to determine the specific gravity test using the gas jar method:

$$\text{Specific Gravity} = \frac{L_2 - L_1}{(L_4 - L_1) - (L_3 - L_2)} \quad (1)$$

Where
L_1: Weight of jar + weight of plates
L_2: The total weight of the sample, plate, and jar, including the plate and the jar itself.
L_3: Water, dry sample, plate, and jar weight together.
L_4: Weight of the jar plus water
The chemical structure of the bottom ash was examined using an X-ray fluorescence analyser. It is performed to assess the effectiveness of extracting phosphates from the furnace's bottom ashes since materials containing calcium, iron, and aluminium can be utilised for phosphate removal. To extract the phosphate in contaminated water, the bottom ashes were employed as a filter. The Box-Behnken design was utilised to improve the technique in terms of burner bottom ash dosage, duration of treatment, and temp. A potassium diphosphate solution in deionised was used to modify the amounts of phosphate in synthetic water. An experiment was conducted in which polluted water was combined for various lengths of time with the furnace's bottom ashes to determine the effect of duration. In such tests, the retention duration ranging from 10 to 70 min, while the burner ash dosages varied from 100 to 1000 mg/L. Calculations of phosphate levels are made with the use of a Hach Lange spectrophotometer. As a result of this study, the Box-Behnken Design was utilised to create an estimation model in order to obtain the highest water treatment. The response variable was phosphorus removal in this research. The determinants were the phosphorus content level, bottom ash dose, and the duration of treatment. In order to construct the estimation matrix, Minitab 20 was chosen. The ranges of predictor factors considered are shown in Table 1.

Table 1. Varying ranges of independent and dependent variables were examined.

Factors	Code	Range
Concentration of phosphate in water	CoF	5-15
The dose of bottom ashes	DoBA	100-1000
Time of the treatment	ToT	10-70

A powerful determinant of linearity and relevance is the α value, which may be calculated using the formula below:

$$\alpha = \sqrt[4]{2X} \quad (2)$$

Where X is the number of variables included in the model.

3. Results and Discussion

The results of the grain size analysis showed the majority of the bottom ash grains are maintained on sieves of lower diameters of less than 16. The results of the sieving tests show that the majority of the bottom ash samples consisted of small particles within 1.180 mm and 4.750mm in size. It was computed using the formulas 3 and 4:

$$\text{Uniformity coefficient} = \frac{x_2}{x_1} \quad (3)$$

$$\text{The gradation coefficient} = \frac{x_3^2}{x_1 * x_2} \quad (4)$$

When plotting the passing rate with particle diameter, x_2, x_3, and x_1 show the percentages of particles that pass through each size. Analysis indicated that x_2, x_3, and x_1 the quantity of the first term were 3.06, 1.36, and 0.48, correspondingly. Furthermore, the uniformity coefficient is 6.21 the gradation coefficient is 1.32. The specimens of bottom ash are judged to be of good quality. Because of this, it is unlikely that the bottom ashes would float, as their relative density is greater than those of the liquid. Permeability was 0.61 m^2/g, while the total area was 6.81 m^2/g. A wide surface, good permeability, and high relative density make the bottoms of the furnace excellent for phosphorus extraction. Chemists determined that the bottom ashes included aluminium, iron, and calcium oxide, magnesium in roughly 25% of the furnace's ash. As a result, it may be argued that the bottom ash of a burner is suitable for removing phosphates. To investigate the effectiveness of burner ash in removing phosphorus compounds from wastewater, batch experiments were run. For each of the 15 tests, the experimental variables were modified in order to identify the optimal experimental set for achieving high removal efficiency at low costs. The findings are shown in Table 2 below. As can be seen in Table 2, the Box-Behnken Design and the removal efficiency obtained are provided.

Table 2. Phosphate elimination in batch flow experiments using a Box-Behnken design.

test	DoBA	CoF	ToT	RE%
01	1000	10	40	90.1
02	1000	10	70	63.1
03	1000	5	40	77.3
04	550	10	40	83.5
05	550	15	70	79.1
06	10	10	10	77.2
07	550	10	40	66.4
08	550	5	10	58.2
09	550	15	10	59.9
10	550	5	70	65.7
11	100	15	40	80.1
12	1000	15	40	79.2
13	100	100	70	77.8
14	100	5	40	79.2
15	550	10	40	70.1

As a result of this investigation, it was shown that the removal efficiencies are proportional to the dose of furnace bottom ashes and the length of time intervals. Phosphorus removal, on the other hand, has an adverse correlation with starting phosphorus concentrations. This is because the accessible surface area of the adsorption is insufficient to collect the phosphorus from the liquid. As a result, there is more surface area accessible for phosphate adsorption from the contaminated water in the furnace bottom ash. It also enhances the interaction of phosphate and the adsorbent. 90.1% of the phosphate concertation in the contaminated water could be removed by using 530 mg/L of bottom ashes for 40 minutes, as shown in Figures 1-3.

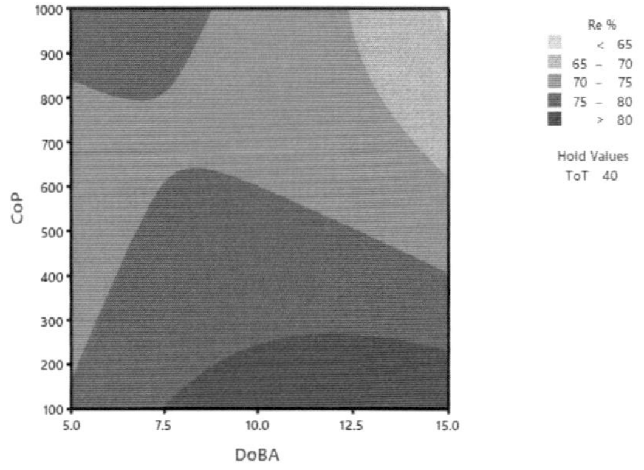

Figure 1: Contour plot of the effects of DoBA and CoP on the removal of phosphate by the bottom ash.

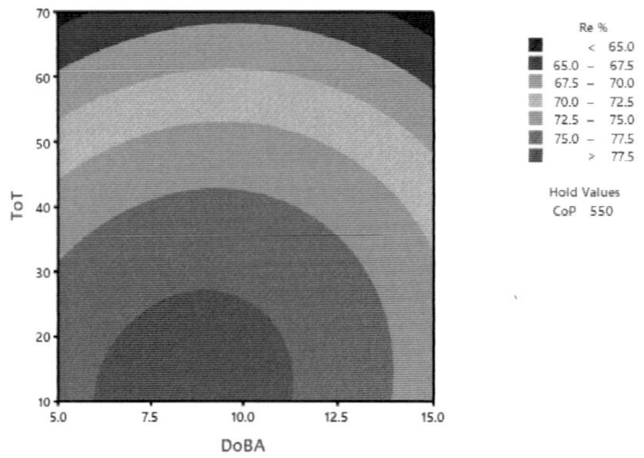

Figure 2: Contour plot of the effects of DoBA and ToT on the removal of phosphate by the bottom ash.

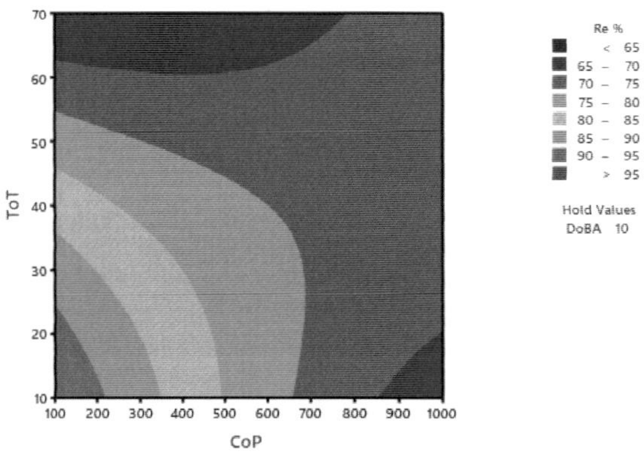

Figure 3: Contour plot of the effects of ToT and CoP on the removal of phosphate by the bottom ash.

To estimate the removal efficiency based on the employed parameters, Box-Behnken Design was used to construct Equation 5.

$$RE\% = 73.33 - 0.44\,DoBA + 2.24\,CoF - 0.10\,ToT + 8.61\,DoBA * DoBA - 2.99\,CoF * CoF - 4.62\,ToT * ToT + 0.25\,DoBA * CoF - 7.17\,DoBA * ToT + 2.93\,CoF * ToT \quad (5)$$

It can be seen from the presented studies about the adsorption method, the depletion of the active sites results in a serious reduction in the adsorption process, therefore; the application of successful sensors (in the literature) helps the operators of the treatment plant to monitor the depletion process, two good examples of the sensors that widely used in the civil and water engineering are the microwave [57-59] and electromagnetic [60-63] sensors.

4. Conclusion

The prospect of using furnace bottom ashes to treat contaminated water for phosphate was studied in this invistigtion. The chemical and physical properties of furnace bottom ashes are essential to remove phosphorus from contaminated water, according to research. This is in line with earlier research findings of about 90 % elimination using the bottom ashes. The dose of the bottom ashes as well as the detention time is positively related to the percentage of phosphorus removal from the polluted wastewater. Relative to phosphorus adsorption capacity, on the other hand, it is inversely correlated with the initial concentration of the pollutant. On the basis of furnace bottom ash dosage, phosphate content, and treatment duration, Box-model Behnken's may also forecast how the removal efficiency will vary.

References

[1] Zubaidi S L, Al-Bugharbee H, Muhsen Y R, Hashim K, Alkhaddar R M, Al-Jumeily D and Aljaaf A J 2019 The Prediction of Municipal Water Demand in Iraq: A Case Study of Baghdad Governorate. In: *12th International Conference on Developments in eSystems Engineering (DeSE),* (Kazan, Russia pp 274-7

[2] Salah Z, Abdulkareem I H, Hashim K S, Al-Bugharbee H, Ridha H M, Gharghan S K, Al-Qaim F F, Muradov M, Kot P and Alkhaddar R 2020 Hybridised Artificial Neural Network model with Slime Mould Algorithm: A novel methodology for prediction urban stochastic water demand *Water* **12** 1-18

[3] Zubaidi S, Al-Bugharbee H, Muhsin Y R, Hashim K and Alkhaddar R 2020 Forecasting of monthly stochastic signal of urban water demand: Baghdad as a case study. In: *IOP Conference Series: Materials Science and Engineering*: IOP Publishing) p 012018

[4] Al-Saati N H, Omran I I, Salman A A, Al-Saati Z and Hashim K S 2021 Statistical modeling of monthly streamflow using time series and artificial neural network models: Hindiya Barrage as a case study *Water Practice and Technology* **16** 681-91

[5] Salah Z, Ortega-Martorell S, Kot P, Alkhaddar R M, Abdellatif M, Gharghan S K, Ahmed M S and Hashim K 2020 A Method for Predicting Long-Term Municipal Water Demands Under Climate Change *Water Resources Management* **34** 1265-79

[6] Salah Z, Hashim K, Ethaib S, Al-Bdairi N S S, Al-Bugharbee H and Gharghan S K 2020 A novel methodology to predict monthly municipal water demand based on weather variables scenario *Journal of King Saud University-Engineering Sciences* **32** 1-18

[7] Zubaidi S, Ortega-Martorell S, Al-Bugharbee H, Olier I, Hashim K S, Gharghan S K, Kot P and Al-Khaddar R 2020 Urban Water Demand Prediction for a City that Suffers from Climate Change and Population Growth: Gauteng Province case study *Water* **12** 1-18

[8] Zubaidi S L, Kot P, Hashim K, Alkhaddar R, Abdellatif M and Muhsin Y R 2019 Using LARS–WG model for prediction of temperature in Columbia City, USA. In: *IOP Conference Series: Materials Science and Engineering*: IOP Publishing) p 012026

[9] Zubaidi Salah L, Al-Bugharbee H, Ortega Martorell S, Gharghan S, Olier I, Hashim K, Al-Bdairi N and Kot P 2020 A Novel Methodology for Prediction Urban Water Demand by

Wavelet Denoising and Adaptive Neuro-Fuzzy Inference System Approach *Water* **12** 1-17

[10] Hashim K S, Al-Saati N H, Alquzweeni S S, Zubaidi S L, Kot P, Kraidi L, Hussein A H, Alkhaddar R, Shaw A and Alwash R 2019 Decolourization of dye solutions by electrocoagulation: an investigation of the effect of operational parameters. In: *First International Conference on Civil and Environmental Engineering Technologies (ICCEET),* (University of Kufa, Iraq pp 25-32

[11] Hashim K S, Hussein A H, Zubaidi S L, Kot P, Kraidi L, Alkhaddar R, Shaw A and Alwash R 2019 Effect of initial pH value on the removal of reactive black dye from water by electrocoagulation (EC) method. In: *2nd International Scientific Conference,* (Al-Qadisiyah University, Iraq pp 12-22

[12] Hashim K S, Khaddar R A, Jasim N, Shaw A, Phipps D, Kot P, Pedrola M O, Alattabi A W, Abdulredha M and Alawsh R 2019 Electrocoagulation as a green technology for phosphate removal from River water *Separation and Purification Technology* **210** 135-44

[13] Abdulredha M, Rafid A, Jordan D and Alattabi A 2017 Facing up to waste: how can hotel managers in Kerbala, Iraq, help the city deal with its waste problem? *Procedia engineering* **196** 771-8

[14] Abdulredha M, Abdulridha A, Shubbar A, Alkhaddar R, Kot P and Jordan D 2020 Estimating municipal solid waste generation from service processions during the Ashura religious event. In: *IOP Conference Series: Materials Science and Engineering*: IOP Publishing) p 012075

[15] Abdulredha M, Kot P, Al Khaddar R, Jordan D and Abdulridha A 2020 Investigating municipal solid waste management system performance during the Arba'een event in the city of Kerbala, Iraq *Environment, Development and Sustainability* **22** 1431-54

[16] Farhan S L, Antón D, Akef V S, Zubaidi S L and Hashim K S 2021 Factors influencing the transformation of Iraqi holy cities: the case of Al-Najaf *Scientific Review Engineering and Environmental Sciences* **30** 365-75

[17] Hashim K S, Al-Saati N H, Hussein A H and Al-Saati Z N 2018 An investigation into the level of heavy metals leaching from canal-dreged sediment: a case study metals leaching from dreged sediment. In: *First International Conference on Materials Engineering & Science,* (Istanbul Aydın University (IAU), Turkey pp 12-22

[18] Hashim K S, Ewadh H M, Muhsin A A, Zubaidi S L, Kot P, Muradov M, Aljefery M and Al-Khaddar R 2020 Phosphate removal from water using bottom ash: Adsorption performance, coexisting anions and modelling studies *Water Science and Technology* **83** 1-17

[19] Alenezi A K, Hasan H A, Hashim K S, Amoako-Attah J, Gkantou M, Muradov M, Kot P and Abdulhadi B 2020 Zeolite-assisted electrocoagulation for remediation of phosphate from calcium-phosphate solution. In: *IOP Conference Series: Materials Science and Engineering*: IOP Publishing) p 012031

[20] Al-Marri S, AlQuzweeni S S, Hashim K S, AlKhaddar R, Kot P, AlKizwini R S, Zubaidi S L and Al-Khafaji Z S 2020 Ultrasonic-Electrocoagulation method for nitrate removal from water. In: *IOP Conference Series: Materials Science and Engineering*: IOP Publishing) p 012073

[21] Mohammed A-H, Hussein A H, Yeboah D, Al Khaddar R, Abdulhadi B, Shubbar A A and Hashim K S 2020 Electrochemical removal of nitrate from wastewater. In: *IOP Conference Series: Materials Science and Engineering*: IOP Publishing) p 012037

[22] Alhendal M, Nasir M J, Hashim K S, Amoako-Attah J, Al-Faluji D, Muradov M, Kot P and Abdulhadi B 2020 Cost-effective hybrid filter for remediation of water from fluoride. In: *IOP Conference Series: Materials Science and Engineering*: IOP Publishing) p 012038

[23] Abdulraheem F S, Al-Khafaji Z S, Hashim K S, Muradov M, Kot P and Shubbar A A 2020 Natural filtration unit for removal of heavy metals from water. In: *IOP Conference Series: Materials Science and Engineering*: IOP Publishing) p 012034

[24] Omran I I, Al-Saati N H, Hashim K S, Al-Saati Z N, Patryk K, Khaddar R A, Al-Jumeily D, Shaw A, Ruddock F and Aljefery M 2019 Assessment of heavy metal pollution in the Great Al-Mussaib irrigation channel *Desalination and Water Treatment* **168** 165-74

[25] Hashim K, Kot P, Zubaid S, Alwash R, Al Khaddar R, Shaw A, Al-Jumeily D and Aljefery M 2020 Energy efficient electrocoagulation using baffle-plates electrodes for efficient Escherichia Coli removal from Wastewater *Journal of Water Process Engineering* **33** 101079-86

[26] Hashim K S, Ali S S M, AlRifaie J K, Kot P, Shaw A, Al Khaddar R, Idowu I and Gkantou M 2020 Escherichia coli inactivation using a hybrid ultrasonic–electrocoagulation reactor *Chemosphere* **247** 125868-75

[27] Alenazi M, Hashim K S, Hassan A A, Muradov M, Kot P and Abdulhadi B 2020 Turbidity removal using natural coagulants derived from the seeds of strychnos potatorum: statistical and experimental approach. In: *IOP Conference Series: Materials Science and Engineering*: IOP Publishing) p 012064

[28] Al-Saati N H, Hussein T K, Abbas M H, Hashim K, Al-Saati Z N, Kot P, Sadique M, Aljefery M H and Carnacina I 2019 Statistical modelling of turbidity removal applied to non-toxic natural coagulants in water treatment: a case study *Desalination and Water Treatment* **150** 406-12

[29] Aqeel K, Mubarak H A, Amoako-Attah J, Abdul-Rahaim L A, Al Khaddar R, Abdellatif M, Al-Janabi A and Hashim K S 2020 Electrochemical removal of brilliant green dye from wastewater. In: *IOP Conference Series: Materials Science and Engineering*: IOP Publishing) p 012036

[30] Alnaimi H, Idan I J, Al-Janabi A, Hashim K, Gkantou M, Zubaidi S L, Kot P and Muradov M 2020 Ultrasonic-electrochemical treatment for effluents of concrete plants. In: *IOP Conference Series Materials Science and Engineering* (University of Kufa, Najaf, Iraq pp 1-9

[31] Alyafei A, AlKizwini R S, Hashim K S, Yeboah D, Gkantou M, Al Khaddar R, Al-Faluji D and Zubaidi S L 2020 Treatment of effluents of construction industry using a combined filtration-electrocoagulation method. In: *IOP Conference Series: Materials Science and Engineering*: IOP Publishing) p 012032

[32] Zanki A K, Mohammad F H, Hashim K S, Muradov M, Kot P, Kareem M M and Abdulhadi B 2020 Removal of organic matter from water using ultrasonic-assisted electrocoagulation method. In: *IOP Conference Series: Materials Science and Engineering*: IOP Publishing) p 012033

[33] Emamjomeh M M, Kakavand S, Jamali H A, Alizadeh S M, Safdari M, Mousavi S E S, Hashim K S and Mousazade M 2020 The treatment of printing and packaging wastewater by electrocoagulation–flotation: the simultaneous efficacy of critical parameters and economics *Desalination and water treatment* **205** 161-74

[34] Abdulhadi B, Kot P, Hashim K, Shaw A, Muradov M and Al-Khaddar R 2021 Continuous-flow electrocoagulation (EC) process for iron removal from water: Experimental, statistical and economic study *Science of The Total Environment* **760** 1-16

[35] Hashim K S, Shaw A, AlKhaddar R, Kot P and Al-Shamma'a A 2021 Water purification from metal ions in the presence of organic matter using electromagnetic radiation-assisted treatment *Journal of Cleaner Production* **280** 1-17

[36] Omran I I, Al-Saati N H, Al-Saati H H, Hashim K S and Al-Saati Z N 2021 Sustainability assessment of wastewater treatment techniques in urban areas of Iraq using multi-criteria decision analysis (MCDA) *Water Practice and Technology* **16** 648-60

[37] Grmasha R A, Al-sareji O J, Salman J M, Hashim K S and Jasim I A 2020 Polycyclic Aromatic Hydrocarbons (PAHs) in Urban Street Dust Within Three Land-Uses of Babylon Governorate, Iraq: Distribution, Sources, and Health Risk Assessment *Journal of King Saud University - Engineering Sciences* **33** 1-18

[38] Al-Sareji O J, Grmasha R A, Salman J M, Idowu I and Hashim K S 2021 Street dust contamination by heavy metals in Babylon governorate, Iraq *Journal of Engineering Science and Technology* **16** 3528 - 46

[39] Chabuk A, Hammood Z A, Abed S A, Kadhim M M, Hashim K, Al-Ansari N and Laue J 2021 Noise Level in Textile Industries: Case Study Al-Hillah Textile Factory-Company for Textile Industries, Al-Hillah-Babylon-Iraq. In: *IOP Conference Series: Earth and Environmental Science*: IOP Publishing) p 012048

[40] Shubbar A A, Al-Shaer A, AlKizwini R S, Hashim K, Hawesah H A and Sadique M 2019 Investigating the influence of cement replacement by high volume of GGBS and PFA on the mechanical performance of cement mortar. In: *First International Conference on Civil and Environmental Engineering Technologies (ICCEET),* (University of Kufa, Iraq pp 31-8

[41] Kadhim A, Sadique M, Al-Mufti R and Hashim K 2020 Developing One-Part Alkali-Activated metakaolin/natural pozzolan Binders using Lime Waste as activation Agent *Advances in Cement Research* **32** 1-38

[42] Majdi H S, Shubbar A, Nasr M S, Al-Khafaji Z S, Jafer H, Abdulredha M, Masoodi Z A, Sadique M and Hashim K 2020 Experimental data on compressive strength and ultrasonic pulse velocity properties of sustainable mortar made with high content of GGBFS and CKD combinations *Data in Brief* **31** 105961-72

[43] Al-Jumeily D, Hashim K, Alkaddar R, Al-Tufaily M and Lunn J 2019 Sustainable and Environmental Friendly Ancient Reed Houses (Inspired by the Past to Motivate the Future). In: *11th International Conference on Developments in eSystems Engineering (DeSE),* (Cambridge, UK pp 214-9

[44] Shubbar A A, Sadique M, Nasr M S, Al-Khafaji Z S and Hashim K S 2020 The impact of grinding time on properties of cement mortar incorporated high volume waste paper sludge ash *Karbala International Journal of Modern Science* **6** 1-23

[45] Shubbar A A, Sadique M, Shanbara H K and Hashim K 2020 *The Development of a New Low Carbon Binder for Construction as an Alternative to Cement. In Advances in Sustainable Construction Materials and Geotechnical Engineering* (Berlin: Springer)

[46] Obaid M K, Nasr M S, Ali I M, Shubbar A A and Hashim K S 2021 Performance of green mortar made from locally available waste tiles and silica fume *Journal of Engineering Science and Technology* **16** 136-51

[47] Emamjomeh M M, Mousazadeh M, Mokhtari N, Jamali H A, Makkiabadi M, Naghdali Z, Hashim K S and Ghanbari R 2020 Simultaneous removal of phenol and linear alkylbenzene sulfonate from automotive service station wastewater: Optimization of coupled electrochemical and physical processes *Separation Science and Technology* **55** 3184-94

[48] Abdulla G, Kareem M M, Hashim K S, Muradov M, Kot P, Mubarak H A, Abdellatif M and Abdulhadi B 2020 Removal of iron from wastewater using a hybrid filter. In: *IOP Conference Series: Materials Science and Engineering*: IOP Publishing) p 012035

[49] Hashim K S, AlKhaddar R, Shaw A, Kot P, Al-Jumeily D, Alwash R and Aljefery M H 2020 *Electrocoagulation as an eco-friendly River water treatment method. In Advances in Water Resources Engineering and Management* (Berline: Springer)

[50] Abdulhadi B A, Kot P, Hashim K S, Shaw A and Khaddar R A 2019 Influence of current density and electrodes spacing on reactive red 120 dye removal from dyed water using electrocoagulation/electroflotation (EC/EF) process. In: *First International*

Conference on Civil and Environmental Engineering Technologies (ICCEET), (University of Kufa, Iraq pp 12-22

[51] Kadhim A, Sadique M, Al-Mufti R and Hashim K 2020 Long-term performance of novel high-calcium one-part alkali-activated cement developed from thermally activated lime kiln dust *Journal of Building Engineering* **32** 1-17

[52] Abdulraheem M S and Kadhum M M 2018 Experimental and numerical study on post-fire behaviour of concentrically loaded reinforced reactive powder concrete columns *Construction and Building Materials* **168** 877-92

[53] Hashim A M and Kadhum M M 2021 Numerical and Experimental Study of Postfire Behavior of Concentrically Loaded SIFCON Columns *ACI Structural Journal* **118** 73-86

[54] Khamees S S, Kadhum M M and Alwash N A 2020 Experimental and numerical investigation on the axial behavior of solid and hollow SIFCON columns *SN Applied Sciences* **2** 1-15

[55] Kadhum M 2015 Prediction of Mechanical Properties of Reactive Powder Concrete by Using Artificial Neural Network and Regression Technique after the Exposure to Fire Flame *Jordan Journal of Civil Engineering* **159** 1-19

[56] Kadhum M M, Alwash N A, Tuama W K and Abdulraheem M S 2019 Experimental and numerical study of influence of crude oil products on the behavior of reactive powder and normal strength concrete slabs *Journal of King Saud University-Engineering Sciences* **32** 293-302

[57] Gkantou M, Muradov M, Kamaris G S, Hashim K, Atherton W and Kot P 2019 Novel Electromagnetic Sensors Embedded in Reinforced Concrete Beams for Crack Detection *Sensors* **19** 5175-89

[58] Ryecroft S P, shaw A, Fergus P, Kot P, Hashim K and Conway L 2019 A Novel Gesomin Detection Method Based on Microwave Spectroscopy. In: *12th International Conference on Developments in eSystems Engineering (DeSE),* (Kazan, Russia pp 429-33

[59] Omer G, Kot P, Atherton W, Muradov M, Gkantou M, Shaw A, Riley M, Hashim K and Al-Shamma'a A 2020 A Non-Destructive Electromagnetic Sensing Technique to Determine Chloride Level in Maritime Concrete *Karbala International Journal of Modern Science* **6** 1-14

[60] Ryecroft S, Shaw A, Fergus P, Kot P, Hashim K, Moody A and Conway L 2019 A First Implementation of Underwater Communications in Raw Water Using the 433 MHz Frequency Combined with a Bowtie Antenna *Sensors* **19** 1813-23

[61] Kot P, Muradov M, Gkantou M, Kamaris G S, Hashim K and Yeboah D 2021 Recent Advancements in Non-Destructive Testing Techniques for Structural Health Monitoring *Applied Sciences* **11** 1-28

[62] Omer G, Kot P, Atherton W, Muradov M, Gkantou M, Shaw A, Riley M, Hashim K and Al-Shamma'a A 2021 A Non-Destructive Electromagnetic Sensing Technique to Determine Chloride Level in Maritime Concrete *Karbala International Journal of Modern Science* **7** 61-71

[63] Ryecroft S, Shaw A, Fergus P, Kot P, Hashim K, Tang A, Moody A and Conway L 2021 An Implementation of a Multi-Hop Underwater Wireless Sensor Network using Bowtie Antenna *Karbala International Journal of Modern Science* **7** 113-29

Recycling of industrial by-products in concrete: Experimental study

R. Hussian[1,*], A. Adnan[1]
[1] Department of Environment, Ministry of Oil, Iraq
Email: rafidbabylon@gmail.com

Abstract. The construction industry consumes enormous amounts of ordinary Portland cement (OPC) to bond the fine and coarse aggregates that give the concrete strength. However, the OPC causes many environmental severe impacts, such as the huge generation of CO_2 gas and alkaline wastewater, which triggered concerns about the health consequences of OPC manufacturing. Therefore, the search for eco-friendly alternatives for OPC has increased rapidly during the last decades, such as the recycling of agricultural and industrial by-products. In this research, residuals of cement furnaces (the dust of cement furnace) and silica fume were mixed and the latter was used as a partial replacement for the OPC. Initially, the chemical structure of the cement dust, silica fume and the OPC were tested. Then, three different (0-45%) percentages of the cement dust and silica fume were added to the concrete mix. The mechanical properties of the produced concrete were investigated using a non-destructive method (ultrasonic pulse velocity) and compared with the properties of the traditional OPC concrete. The results showed the cement dust and silica fume contain the favourable oxides, and the best mixing ratio was 12.5% of silica fume, 12.5% of cement dust and 75% of OPC that achieved pulse velocity of 4102 m/s (the best compressive strength).

Keywords: ultrasonic pulse velocity, cement kiln dust, silica fume.

1. Introduction

It is well-known huge quantities of Ordinary Portland Cement (OPC) are consumed in the construction of the infrastructure of cities, such as the construction bridges, houses, high rise buildings, pavements, and sewers [1, 2]. Although is a proved fact that OPC plays a major role in the development of civilisation and the global financial system, its role in environmental deterioration is also proved [3-5]. OPC factories are estimated to emit about 10% of total CO_2 in the air, leading to catastrophic environmental effects such as weather change [6, 7]. Therefore, the pollution and consumption of water have risen substantially [8, 9]. The released effluents from concrete plants and casting activities are highly particular due to the chemical composition of Conventional concrete [10, 11]. Also, the production of solid waste from cement industry was has resin remarkably [12, 13], especially from the demolition of old concrete structures in the cities (municipal areas) [14, 15]. As a result, devastating impacts on the overall quality of water and the degradation of living beings have occurred [16, 17]. Different treatments techniques have been developed to remove many pollutants found in cement plant effluent and other industrial wastewaters [18-20], including filtrations [21-24], coagulation [25-29], electrocoagulation [30-32], sonication-assisted [33, 34], electro-chemical [35-37], electro-physical [38-41], and hybridised methods [42-45]. However, recent studies show an increase in water pollution [39-41] and freshwater consumption [46-48]. Additionally, these techniques are inadequate to control all anticipated pollutants from cement factories. Because of the huge volumes of Carbon dioxide released into the air from the operations related to conventional cement, as well as contaminants discharged into

water bodies, cement manufacture has become a growing subject of attention [49-51]. As a consequence, reviews show that the development of replacements for cement components, such as silica fume (SF) and cement kiln dust (CKD), is the most viable solution to minimise cement production [52, 53].
Although the utilisation of SF and CKD can improve the properties of concrete, previous studies have proven that the use of these materials has also detrimental effects on the characteristics of concrete. As the percentage of CKD to cement increased, the workability of fresh mixes dropped. Furthermore, as the percentage of CKD increases, the strength of concrete lowers. SCMS materials like SF and CKD are extremely effective and widely employed as cementitious materials because they have large surface surfaces and contain a considerable quantity of silica oxides. Earlier research has shown that replacing Cement with SF at a proportion of (0.22-.30) is an effective way to maintain the strength of concrete. The ultrasonic pulse velocity (UPV) is a non-destructive testing method that is used to assess the quality of concrete buildings. This technique, which involves measuring the velocity of the ultrasonic pulse velocity travelling through concrete members, is used to analyse several attributes of concrete, such as quality and strength [17, 54].
The present invistgation evalutes the effectiveness of SF and CKD as cement substitutes. The main goal of this study is to investigate if these chemicals impact cement characteristics like strength at different curing periods using an ultrasonic pulse velocity test (one and 4) weeks. The ultrsonic pulse technology was used in this study because it cheap and accurate [55-57], and also it could be used on the surface of the concrete sample or by embading it in the body of the concrete and connect it to the reciever using wireless technologies [56, 58-60].

2. Experimentation curriculum

The experimental studies focused on the measurement of the ultrasonic pulse velocity of the concrete mix produced by partially substituting cement with SF and CKD to evaluate its mechanical properties

2.1. Materials

Cement Kiln Dust (CKD) is a by-product of the cement industry. It's a finely powdered substance that looks like Portland cement. usually, CKD is made up of micron-sized grains recovered by combustion processes during the cement clinker manufacturing process [4, 52]. Cement kiln dust CKD is a fine powdery substance that ranges in hue from grey to brown and is relatively homogeneous in dimension. The manufacturing process, dust collecting technique, chemical properties of CKD, and alkali concentration all influence the gradation of CKD. With fly ash and GGBS in various percentages up to 16%, this product could be used as a cementitious substance. If CKD is utilised separately, the resultant combination may have decreased workability, weight, and setting time due to the high alkali concentration [52, 54].

Silica fume is regarded as a by-product of the silicon and ferrosilicon alloy manufacturing industries, and it is produced at extreme temps from quartz reduction. Because of its properties that promote the cementitious reaction, silica fume is widely utilised as a cementitious material in concrete. It is an ultrafine powder consist of 84-96 non-crystalline silica and about 76 percent silicon [4]. The quantity of Silicon dioxide in silica fume is proportional to the kind of alloy generated in the manufacturing facility.
SF partials are extremely tiny and round, roughly 100 times finer than ordinary cement particles. Previous researches show that the SF concrete have reduced bleeding, porosity, and permeability Because SF oxides react with and consume Calcium hydroxide, which is CH generated during cement hydration.
The main binding ingredient in this experiment was Portland Cement, which has strong mechanical characteristics that help the combination to remain coherent. The cement properties used in this study were measured according to BS EN 196-2:2013.
Diagrams 1,2, and 3 show the chemical composition of SF, CKD, and Portland cement. These features meet the requirements of the BS-EN-197-1(2011) and BS-EN-450-1 standards (2012).
The particle size of the grains, and even the chloride and sulfate concentrations, were checked using the BS EN 12620:2002+A1 standard (2008).
Concrete was prepared and treated using impure and organic-free portable water.

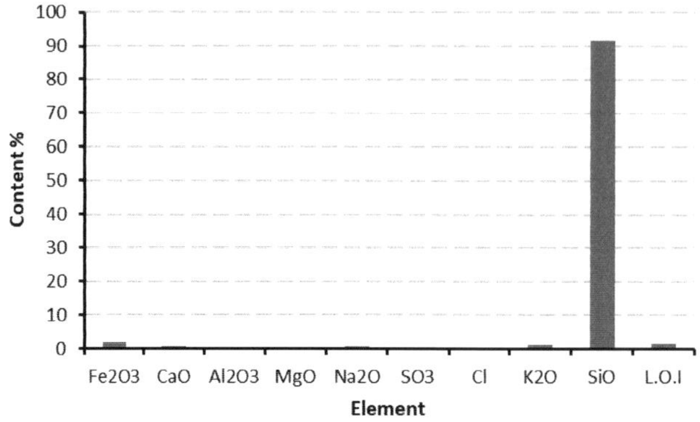

Figure 1: The chemical structure of silica fume.

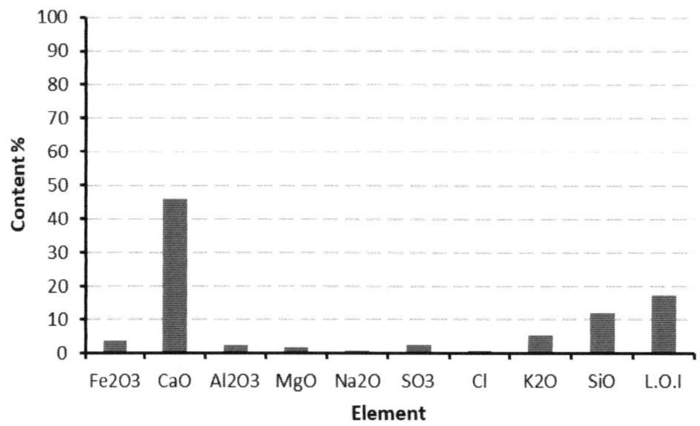

Figure 2: The chemical structure of CKD.

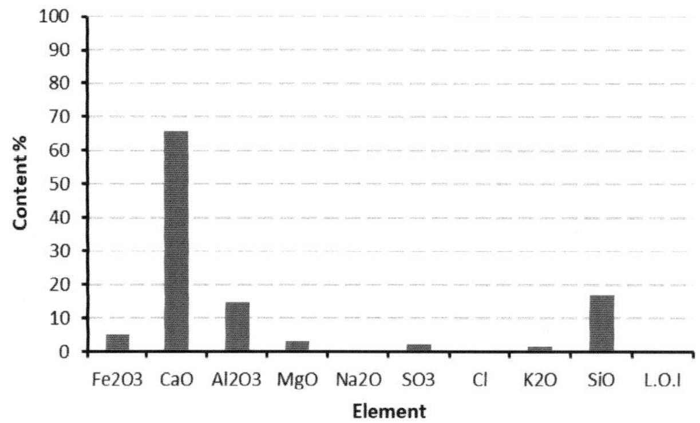

Figure 3: The chemical structure of OPC.

2.2. Testing Techniques

For every mix, three prisms (160x40x40 mm) were cast to see how substituting cement with CKD and SF affected the quality of the mortar mixture. These tests are limited to examining items poured from cement mortar. The investigated specimens are then built up of cement in three different concentrations of CKD and SF . Upon completion of the initial setting time of the mixtures, all samples were maintained in good condition, molded, and placed in water for the cure. BS EN 12504-4:2004 was used to conduct ultrasonic tests on hardening specimens at one week and 4 weeks.

2.3. Design of Mixture

In this study, part of the design process includes determining the proportions of fine aggregate, water, cement, and materials additive ingredients for the control concrete mixes. To match conventional rating curves, fine aggregates have been utilised in the mixture design. The water to binder ratio was 0.4 in all of the mixes. The ratio of sand to a binder that used in this study was 2.4 in all of the mixes. The percentage of each component of the mix is shown in figure (4).

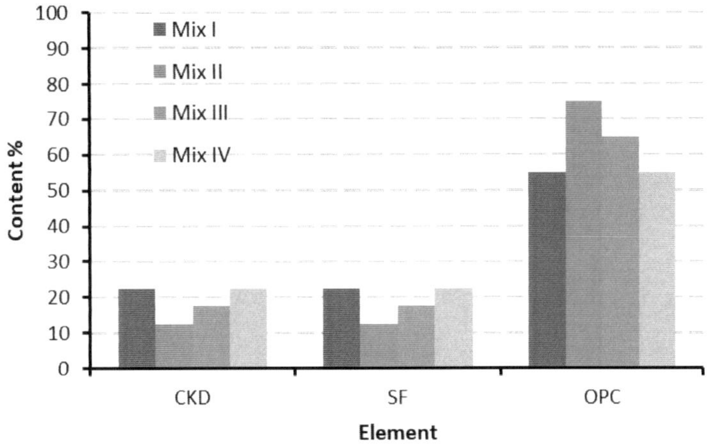

Figure 4: The mix design.

The ultrasonic pulses are sent through the sample to be examined, and the time required for the pulse to permeate the specimen is measured. A high speed implies that the examined structure is of top condition, whereas a low velocity suggests that the examined structure is of bad condition. Pulse producers, a transducer for converting electronic pulses into mechanical pulses with vibrations of 40 to 50 kHz, and a pulses detector are all used in UPV assessment. The velocities of pulses are determined as follows:

$$\text{Pulse velosity} = \frac{\text{The specimen's thickness}}{\text{The required time for the pulse to penetrate the sample}} \quad (1)$$

3. Results

The results of examining a control concrete mix with substitute material of cement with varying quantities of SF and CKD at different curing periods are explained in Table 1 and Figure 5.

Table 1. One week and 4 weeks ultrasonic pulse velocity testing.

Number of tests	Cement (%)	SF (%)	CKD (%)	UPV test (m/s)	
				One week	Four weeks
MIX1	100	0	0	3828	4008
MIX2	75	12.5	12.5	3850	4102
MIX3	65	17.5	17.5	3785	4055
MIX4	55	22.5	22.5	3578	3863

The main conclusions that may be drawn from the findings of this study are listed below:
The use of a partial replacement of SF and CKD in mixes has been demonstrated to lower concrete pulse velocity values by a tiny portion. In comparison to the control mixture, the pulse velocity of mixtures two and three have been enhanced by 2.3% and 11% after 4 weeks of curing, respectively. Whereas, mixture number four decrease the velocity by 3.6 %. This would be based on the view that just a little amount of cementitious materials is required to fill empty fields in the mortar, therefore improving its mechanical properties. Previous studies showed that CKD and SF are ineffectual substances at first and require time to connect with cement components. Therefore, it can be noted that After one week of curing, using 45 percent of a partial substitute decreases the pulse velocity measurements of the mix by around 6.5 percent. while this value fell to 3.6 after curing for 4 weeks. This is because extra cementitious ingredients reduce concrete compressive strength, which is a key component of the manufacture gel (C-S-H) in concrete. After sitting time, they interact and utilise the moisture components, Ca (OH)2, to allow and initiate hydration of silica fume and cement kiln dust.

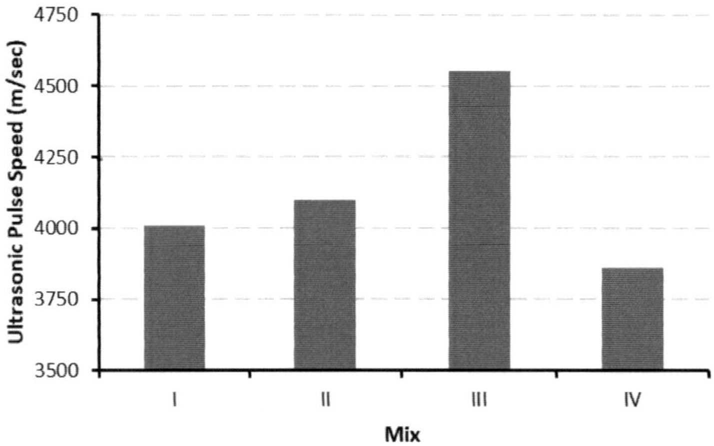

Figure 5. Ultrasonic Pulse Velocity Monitoring with one week and 4weeks curing day mixtures.

According to the observations, specimens that were treated for 4 weeks had greater pulse velocity readings. That was based on the fact that the curing time improves C-S-H, which leads to a decline in the number of interior gaps or porosity in the conventional concrete, which impacts the properties of concrete and enhances its capacity to withstand compressive stresses.
The Ultrasonic Pulse Velocity method, a non-destructive test methodology, was used in this investigation. As a result, more sophisticated procedures for verifying concrete properties are now

accessible. Sensors were used to monitor for microcracks, concrete humidity, and other applications in the past. Additional research might utilise the same approach.

4. Conclusions

The outcomes of the current investigation confirmed the partial replacement of the OPC by silica fume and CKD in concrete increases, the pulse velocity values decrease. The material quality, on the other hand, shows a little improvement with a restricted replacement rate. Whenever the cement in a combination is replaced with extra material, a longer curing period results in a higher-quality specimen. The use of 25 ~ 35 percent additional cementitious material as a cement substitution for cement could be appropriate proportions, with an increment in this ratio resulting in a slight improvement in quality assessment. Moreover, the used strategy in this research was UPV, which is a traditional instrument; consequently, more current technologies, such as high frequency radars and laser, are recommended to evaluate the mechanical properties of concrete.

References

[1] Majdi H S, Shubbar A, Nasr M S, Al-Khafaji Z S, Jafer H, Abdulredha M, Masoodi Z A, Sadique M and Hashim K 2020 Experimental data on compressive strength and ultrasonic pulse velocity properties of sustainable mortar made with high content of GGBFS and CKD combinations *Data in Brief* **31** 105961-72

[2] Farhan S L, Jasim I A and Al-Mamoori S K 2019 The transformation of the city of Najaf, Iraq: Analysis, reality and future prospects *Journal of Urban Regeneration & Renewal* **13** 160-71

[3] Shubbar A A, Sadique M, Nasr M S, Al-Khafaji Z S and Hashim K S 2020 The impact of grinding time on properties of cement mortar incorporated high volume waste paper sludge ash *Karbala International Journal of Modern Science* **6** 1-23

[4] Shubbar A A, Sadique M, Shanbara H K and Hashim K 2020 *The Development of a New Low Carbon Binder for Construction as an Alternative to Cement. In Advances in Sustainable Construction Materials and Geotechnical Engineering* (Berlin: Springer)

[5] Farhan S L, Antón D, Akef V S, Zubaidi S L and Hashim K S 2021 Factors influencing the transformation of Iraqi holy cities: the case of Al-Najaf *Scientific Review Engineering and Environmental Sciences* **30** 365-75

[6] Grmasha R A, Al-sareji O J, Salman J M, Hashim K S and Jasim I A 2020 Polycyclic Aromatic Hydrocarbons (PAHs) in Urban Street Dust Within Three Land-Uses of Babylon Governorate, Iraq: Distribution, Sources, and Health Risk Assessment *Journal of King Saud University - Engineering Sciences* **33** 1-18

[7] Al-Sareji O J, Grmasha R A, Salman J M, Idowu I and Hashim K S 2021 Street dust contamination by heavy metals in Babylon governorate, Iraq *Journal of Engineering Science and Technology* **16** 3528 - 46

[8] Obaid M K, Nasr M S, Ali I M, Shubbar A A and Hashim K S 2021 Performance of green mortar made from locally available waste tiles and silica fume *Journal of Engineering Science and Technology* **16** 136-51

[9] Al-Jumeily D, Hashim K, Alkaddar R, Al-Tufaily M and Lunn J 2019 Sustainable and Environmental Friendly Ancient Reed Houses (Inspired by the Past to Motivate the Future). In: *11th International Conference on Developments in eSystems Engineering (DeSE),* (Cambridge, UK pp 214-9

[10] Farhan S L and Nasar A Z 2021 Urban Identity in the Pilgrimage Cities of Iraq: Analysis Trends of Architectural Designers in the City of Karbala *J. Urban Regen. Renew* **14** 2-14

[11] Jasim I A, Farhan S L, Al-Maliki L A and AL-Mamoori S K 2021 Climatic Treatments for Housing in the Traditional Holy Cities: A Comparison between Najaf and Yazd Cities. In: *IOP Conference Series: Earth and Environmental Science*: IOP Publishing) p 012017

[12] Alyafei A, AlKizwini R S, Hashim K S, Yeboah D, Gkantou M, Al Khaddar R, Al-Faluji D and Zubaidi S L 2020 Treatment of effluents of construction industry using a combined filtration-electrocoagulation method. In: *IOP Conference Series: Materials Science and Engineering*: IOP Publishing) p 012032

[13] Alnaimi H, Idan I J, Al-Janabi A, Hashim K, Gkantou M, Zubaidi S L, Kot P and Muradov M 2020 Ultrasonic-electrochemical treatment for effluents of concrete plants. In: *IOP Conference Series Materials Science and Engineering* (University of Kufa, Najaf, Iraq pp 1-9
[14] Abdulredha M, Abdulridha A, Shubbar A, Alkhaddar R, Kot P and Jordan D 2020 Estimating municipal solid waste generation from service processions during the Ashura religious event. In: *IOP Conference Series: Materials Science and Engineering*: IOP Publishing) p 012075
[15] Abdulredha M, Kot P, Al Khaddar R, Jordan D and Abdulridha A 2020 Investigating municipal solid waste management system performance during the Arba'een event in the city of Kerbala, Iraq *Environment, Development and Sustainability* **22** 1431-54
[16] Abdulredha M, Rafid A, Jordan D and Alattabi A 2017 Facing up to waste: how can hotel managers in Kerbala, Iraq, help the city deal with its waste problem? *Procedia engineering* **196** 771-8
[17] Khamees S S, Kadhum M M and Alwash N A 2020 Experimental and numerical investigation on the axial behavior of solid and hollow SIFCON columns *SN Applied Sciences* **2** 1-15
[18] Hashim K S, Al-Saati N H, Hussein A H and Al-Saati Z N 2018 An investigation into the level of heavy metals leaching from canal-dreged sediment: a case study metals leaching from dreged sediment. In: *First International Conference on Materials Engineering & Science,* (Istanbul Aydın University (IAU), Turkey pp 12-22
[19] Hashim K S, Al-Saati N H, Alquzweeni S S, Zubaidi S L, Kot P, Kraidi L, Hussein A H, Alkhaddar R, Shaw A and Alwash R 2019 Decolourization of dye solutions by electrocoagulation: an investigation of the effect of operational parameters. In: *First International Conference on Civil and Environmental Engineering Technologies (ICCEET),* (University of Kufa, Iraq pp 25-32
[20] Alattabi A W, Harris C, Alkhaddar R, Alzeyadi A and Abdulredha M 2017 Online Monitoring of a sequencing batch reactor treating domestic wastewater *Procedia engineering* **196** 800-7
[21] Abdulraheem F S, Al-Khafaji Z S, Hashim K S, Muradov M, Kot P and Shubbar A A 2020 Natural filtration unit for removal of heavy metals from water. In: *IOP Conference Series: Materials Science and Engineering*: IOP Publishing) p 012034
[22] Alenazi M, Hashim K S, Hassan A A, Muradov M, Kot P and Abdulhadi B 2020 Turbidity removal using natural coagulants derived from the seeds of strychnos potatorum: statistical and experimental approach. In: *IOP Conference Series: Materials Science and Engineering*: IOP Publishing) p 012064
[23] Hashim K S, Ewadh H M, Muhsin A A, Zubaidi S L, Kot P, Muradov M, Aljefery M and Al-Khaddar R 2020 Phosphate removal from water using bottom ash: Adsorption performance, coexisting anions and modelling studies *Water Science and Technology* **83** 1-17
[24] Al-Saati N H, Hussein T K, Abbas M H, Hashim K, Al-Saati Z N, Kot P, Sadique M, Aljefery M H and Carnacina I 2019 Statistical modelling of turbidity removal applied to non-toxic natural coagulants in water treatment: a case study *Desalination and Water Treatment* **150** 406-12
[25] Omran I I, Al-Saati N H, Al-Saati H H, Hashim K S and Al-Saati Z N 2021 Sustainability assessment of wastewater treatment techniques in urban areas of Iraq using multi-criteria decision analysis (MCDA) *Water Practice and Technology* **16** 648-60
[26] Omran I I, Al-Saati N H, Hashim K S, Al-Saati Z N, Patryk K, Khaddar R A, Al-Jumeily D, Shaw A, Ruddock F and Aljefery M 2019 Assessment of heavy metal pollution in the Great Al-Mussaib irrigation channel *Desalination and Water Treatment* **168** 165-74
[27] Ryecroft S P, shaw A, Fergus P, Kot P, Hashim K and Conway L 2019 A Novel Gesomin Detection Method Based on Microwave Spectroscopy. In: *12th International Conference on Developments in eSystems Engineering (DeSE),* (Kazan, Russia pp 429-33
[28] Zubaidi S L, Al-Bugharbee H, Muhsen Y R, Hashim K, Alkhaddar R M, Al-Jumeily D and Aljaaf A J 2019 The Prediction of Municipal Water Demand in Iraq: A Case Study of Baghdad Governorate. In: *12th International Conference on Developments in eSystems Engineering (DeSE),* (Kazan, Russia pp 274-7
[29] Mousazadeh M, Paital B, Naghdali Z, Mortezania Z, Hashemi M, Karamati Niaragh E, Aghababaei M, Ghorbankhani M, Lichtfouse E, Sillanpää M, Hashim K S and Emamjomeh M

M 2021 Positive environmental effects of the coronavirus 2020 episode: a review *Environment, Development and Sustainability* **21** 1-23

[30] Hashim K S, Hussein A H, Zubaidi S L, Kot P, Kraidi L, Alkhaddar R, Shaw A and Alwash R 2019 Effect of initial pH value on the removal of reactive black dye from water by electrocoagulation (EC) method. In: *2nd International Scientific Conference,* (Al-Qadisiyah University, Iraq pp 12-22

[31] Hashim K, Kot P, Zubaid S, Alwash R, Al Khaddar R, Shaw A, Al-Jumeily D and Aljefery M 2020 Energy efficient electrocoagulation using baffle-plates electrodes for efficient Escherichia Coli removal from Wastewater *Journal of Water Process Engineering* **33** 101079-86

[32] Hashim K S, AlKhaddar R, Shaw A, Kot P, Al-Jumeily D, Alwash R and Aljefery M H 2020 *Electrocoagulation as an eco-friendly River water treatment method. In Advances in Water Resources Engineering and Management* (Berline: Springer)

[33] Zanki A K, Mohammad F H, Hashim K S, Muradov M, Kot P, Kareem M M and Abdulhadi B 2020 Removal of organic matter from water using ultrasonic-assisted electrocoagulation method. In: *IOP Conference Series: Materials Science and Engineering*: IOP Publishing) p 012033

[34] Hashim K S, Shaw A, AlKhaddar R, Kot P and Al-Shamma'a A 2021 Water purification from metal ions in the presence of organic matter using electromagnetic radiation-assisted treatment *Journal of Cleaner Production* **280** 1-17

[35] Aqeel K, Mubarak H A, Amoako-Attah J, Abdul-Rahaim L A, Al Khaddar R, Abdellatif M, Al-Janabi A and Hashim K S 2020 Electrochemical removal of brilliant green dye from wastewater. In: *IOP Conference Series: Materials Science and Engineering*: IOP Publishing) p 012036

[36] Mohammed A-H, Hussein A H, Yeboah D, Al Khaddar R, Abdulhadi B, Shubbar A A and Hashim K S 2020 Electrochemical removal of nitrate from wastewater. In: *IOP Conference Series: Materials Science and Engineering*: IOP Publishing) p 012037

[37] Abdulhadi B, Kot P, Hashim K, Shaw A, Muradov M and Al-Khaddar R 2021 Continuous-flow electrocoagulation (EC) process for iron removal from water: Experimental, statistical and economic study *Science of The Total Environment* **760** 1-16

[38] Emamjomeh M M, Kakavand S, Jamali H A, Alizadeh S M, Safdari M, Mousavi S E S, Hashim K S and Mousazade M 2020 The treatment of printing and packaging wastewater by electrocoagulation–flotation: the simultaneous efficacy of critical parameters and economics *Desalination and water treatment* **205** 161-74

[39] Emamjomeh M M, Mousazadeh M, Mokhtari N, Jamali H A, Makkiabadi M, Naghdali Z, Hashim K S and Ghanbari R 2020 Simultaneous removal of phenol and linear alkylbenzene sulfonate from automotive service station wastewater: Optimization of coupled electrochemical and physical processes *Separation Science and Technology* **55** 3184-94

[40] Hashim K S, Ali S S M, AlRifaie J K, Kot P, Shaw A, Al Khaddar R, Idowu I and Gkantou M 2020 Escherichia coli inactivation using a hybrid ultrasonic–electrocoagulation reactor *Chemosphere* **247** 125868-75

[41] Abdulhadi B A, Kot P, Hashim K S, Shaw A and Khaddar R A 2019 Influence of current density and electrodes spacing on reactive red 120 dye removal from dyed water using electrocoagulation/electroflotation (EC/EF) process. In: *First International Conference on Civil and Environmental Engineering Technologies (ICCEET),* (University of Kufa, Iraq pp 12-22

[42] Abdulla G, Kareem M M, Hashim K S, Muradov M, Kot P, Mubarak H A, Abdellatif M and Abdulhadi B 2020 Removal of iron from wastewater using a hybrid filter. In: *IOP Conference Series: Materials Science and Engineering*: IOP Publishing) p 012035

[43] Alenezi A K, Hasan H A, Hashim K S, Amoako-Attah J, Gkantou M, Muradov M, Kot P and Abdulhadi B 2020 Zeolite-assisted electrocoagulation for remediation of phosphate from calcium-phosphate solution. In: *IOP Conference Series: Materials Science and Engineering*: IOP Publishing) p 012031

[44] Alhendal M, Nasir M J, Hashim K S, Amoako-Attah J, Al-Faluji D, Muradov M, Kot P and Abdulhadi B 2020 Cost-effective hybrid filter for remediation of water from fluoride. In: *IOP Conference Series: Materials Science and Engineering*: IOP Publishing) p 012038

[45] Al-Marri S, AlQuzweeni S S, Hashim K S, AlKhaddar R, Kot P, AlKizwini R S, Zubaidi S L and Al-Khafaji Z S 2020 Ultrasonic-Electrocoagulation method for nitrate removal from water. In: *IOP Conference Series: Materials Science and Engineering*: IOP Publishing) p 012073

[46] Salah Z, Abdulkareem I H, Hashim K S, Al-Bugharbee H, Ridha H M, Gharghan S K, Al-Qaim F F, Muradov M, Kot P and Alkhaddar R 2020 Hybridised Artificial Neural Network model with Slime Mould Algorithm: A novel methodology for prediction urban stochastic water demand *Water* **12** 1-18

[47] Salah Z, Hashim K, Ethaib S, Al-Bdairi N S S, Al-Bugharbee H and Gharghan S K 2020 A novel methodology to predict monthly municipal water demand based on weather variables scenario *Journal of King Saud University-Engineering Sciences* **32** 1-18

[48] Salah Z, Ortega-Martorell S, Kot P, Alkhaddar R M, Abdellatif M, Gharghan S K, Ahmed M S and Hashim K 2020 A Method for Predicting Long-Term Municipal Water Demands Under Climate Change *Water Resources Management* **34** 1265-79

[49] Shubbar A A, Al-Shaer A, AlKizwini R S, Hashim K, Hawesah H A and Sadique M 2019 Investigating the influence of cement replacement by high volume of GGBS and PFA on the mechanical performance of cement mortar. In: *First International Conference on Civil and Environmental Engineering Technologies (ICCEET),* (University of Kufa, Iraq pp 31-8

[50] Hashim A M and Kadhum M M 2021 Numerical and Experimental Study of Postfire Behavior of Concentrically Loaded SIFCON Columns *ACI Structural Journal* **118** 73-86

[51] Kadhum M 2015 Prediction of Mechanical Properties of Reactive Powder Concrete by Using Artificial Neural Network and Regression Technique after the Exposure to Fire Flame *Jordan Journal of Civil Engineering* **159** 1-19

[52] Kadhim A, Sadique M, Al-Mufti R and Hashim K 2020 Long-term performance of novel high-calcium one-part alkali-activated cement developed from thermally activated lime kiln dust *Journal of Building Engineering* **32** 1-17

[53] Kadhum M M, Alwash N A, Tuama W K and Abdulraheem M S 2019 Experimental and numerical study of influence of crude oil products on the behavior of reactive powder and normal strength concrete slabs *Journal of King Saud University-Engineering Sciences* **32** 293-302

[54] Kadhim A, Sadique M, Al-Mufti R and Hashim K 2020 Developing One-Part Alkali-Activated metakaolin/natural pozzolan Binders using Lime Waste as activation Agent *Advances in Cement Research* **32** 1-38

[55] Omer G, Kot P, Atherton W, Muradov M, Gkantou M, Shaw A, Riley M, Hashim K and Al-Shamma'a A 2020 A Non-Destructive Electromagnetic Sensing Technique to Determine Chloride Level in Maritime Concrete *Karbala International Journal of Modern Science* **6** 1-14

[56] Kot P, Muradov M, Gkantou M, Kamaris G S, Hashim K and Yeboah D 2021 Recent Advancements in Non-Destructive Testing Techniques for Structural Health Monitoring *Applied Sciences* **11** 1-28

[57] Omer G, Kot P, Atherton W, Muradov M, Gkantou M, Shaw A, Riley M, Hashim K and Al-Shamma'a A 2021 A Non-Destructive Electromagnetic Sensing Technique to Determine Chloride Level in Maritime Concrete *Karbala International Journal of Modern Science* **7** 61-71

[58] Ryecroft S, Shaw A, Fergus P, Kot P, Hashim K, Tang A, Moody A and Conway L 2021 An Implementation of a Multi-Hop Underwater Wireless Sensor Network using Bowtie Antenna *Karbala International Journal of Modern Science* **7** 113-29

[59] Ghazali R, Hussain A J, Al-Jumeily D and Merabti M 2007 Dynamic ridge polynomial neural networks in exchange rates time series forecasting. In: *International Conference on Adaptive and Natural Computing Algorithms*: Springer) pp 123-32

[60] Fergus P, Hussain A, Al-Jumeily D, Huang D-S and Bouguila N 2017 Classification of caesarean section and normal vaginal deliveries using foetal heart rate signals and advanced machine learning algorithms *Biomedical engineering online* **16** 1-26

Sono-assisted treatment of textile wastewater: reactive black 5 dye a case study

R. S. Alwash[1,*], Safaa K.H[2], Abduljaleel Al-Janabi[2]

[1] Environmental Engineering Department, University of Technology, Iraq
[2] Engineering Faculty, University of Babylon, Babel, Iraq

Corresponding author's e-mail: rehamalwosh@yahoo.com

Abstract. The textile industry provides the global jobs market with a considerable number of jobs every year, and it supports the national economy of many countries, such as India, Turkey, Canada and many European countries. However, textile wastewater is very harmful to the environment and the ecosystem. The reports refer to the toxic effects of the textile industry on aquatic life, human, and also on the quality of water and soils. Also, the volumes of textile wastewater have dramatically increased over the last years because of the increase in the global demand for fabrics and clothes. Therefore, the need for efficient treatment became more urgent than ever before. The present work focuses on the use of the sonication process to enhance the electro-coagulation units' (EC) ability to remove dyes from textile wastewater. Synthetic water containing 28 mg/L of reactive black 5 dye was treated firstly in the electro-coagulation for 1 hour at initial pH of 4, 5, 6, 7, and 8, and current density of 1, 2 and 3 mA/cm^2. Then, the wastewater was sonicated for 10 minutes before the electro-coagulation process. The results showed the lowest pH (4), and the highest current density (2 mA/cm^2) achieved the highest removal efficiency in the EC unit (82.14%), and the use of the sonication process before the EC unit increased the removal of the dye to 100%.

1. Introduction

The phenomenon of water pollution is increasing nowadays because of the increase in the generation of wastewater from different industrial sectors, agriculture, and the domestic use of water [1-3]. For example, the reports referred to a dramatic increase in the usage of freshwater over the last few decades due to the rise in the world population that recently reached 7 billion people [4-6], this increase in the world population is companied by the increase in the industrial and agricultural activities to meet the daily demands for the goods and food of this massive number of people [7, 8].

For example, the global statistics show the consumed amount of dyes is 700,000 tonnes per year, and the required volume of freshwater in the textile industry to produce 1000 g of fabric is 100 to 200 litres [9, 10]. This type of wastewater disposal to the rivers or lakes has potential severe effects on humans and aquatic life. For example, coloured wastewater limits the sunlight; consequently, many forms of aquatic life will disappear due to the lack of nutrients. Moreover, the dyes, especially azo dyes, in water undergo a series of reactions that lead to the production of carcinogenic agents, which cause serious health issues for humans [11, 12]. Moreover, these chemical reactions result in rapid consumption of the dissolved oxygen, and consequently, it leads to the mass death of fish, plants and other forms of life, which negatively influences the food chain [13, 14].

Hence, a broad range of treatments was used to remove dyes from the textile wastewater before the final disposal to the water bodies [12, 15], such as aerobic and anaerobic degradations, adsorptions, and filtrations [16-18]. Inappropriately, most conventional treatments are not effective in the removal of

dyes due to the complexity of textile wastewater, and also, some of the conventional treatments are too expensive to be used in developing countries. Therefore, the new trend in the researches is the use of a combination of methods to treat complex wastewaters, such as the combinations of filtration-electrocoagulation and ultrasonic-coagulation [19-21].
In this paper, the sono-electrocoagulation method will be used to remove a textile dye from water, which is the reactive black 5 dye, which was used as a model dye because of its wide use in the textile industry [10, 11].

2. Methodology

The reactive black 5 dye was purchased from Sigma Aldrich and used without further processing. The dye was dissolved in distilled water to produce a stock solution with initial concentrations of 100 mg/L, which was used to make diluted water samples with a concentration of 28 mg/L that were used in the experiments.
The experiments were conducted using an electro-coagulation unit, shown in Fig. 1. The unit is made from transparent plastic with a working volume of 1 litre and has 4 plates (electrodes). The electrodes are aluminium with an area of 500 cm^2. The dyed solution was circulated using a peristaltic water pump with a maximum capacity of 5 L/ hour. The electrolysing process was done using a rectifier (SP-3010). The sonication process was done using a 2.7 L ultrasonic bath (FB15051) with variable temperatures (up to 60 °C) and powers (up to 280W), and a constant frequency (37000 Hz). The experimental work was according to the literature [22-25]

Figure 1. The electro-coagulation unit.

The treatment was done in two phases; the first phase was conducted using the electro-coagulation unit separately. Water samples containing 28 mg/L of reactive black 5 dye were first treated in the electro-coagulation for 1 hour at an initial pH of 4, 5, 6, 7, and 8, with a current density of 1, 2 and 3 mA/cm^2. The removal efficiency was recorded after each run.
The second phase was done using both sonication and electro-coagulation; water samples were sonicated for 10.0 minutes before the electrolysing process, then treated in the electro-coagulation unit at the best conditions from phase 1.
The removal efficiency was calculated as follows [26, 27]:

$$\text{Percentage of dye removal} = \frac{Initial\ dye\ concentration - measured\ dye\ concentration}{Initial\ dye\ concentration} \times 100 \quad (1)$$

The concentration of dyes was spectrophotometrically measured using a Hach Lange device. Solids were separated from water samples, pre-measurement of concentration, using paper filters (0.45 μm) [28-30]

IOP Conf. Series: Materials Science and Engineering 1184 (2021) 012021 doi:10.1088/1757-899X/1184/1/012021

3. Results

The results of the first phase of this paper were obtained from the electro-coagulation experiments, as follows:

3.1. Effects of the pH

The pH is an important factor in the electro-coagulation process because it affects the sizes of the gas bubbles that affect the floatation of pollutants (the removal of pollutants) and also the chemical form of the aluminium coagulants depends on the pH of the solution [31-33]. Therefore, studying the effect of pH on the removal of pollutants is an important part of electro-coagulation studies.

In this paper, the effects of the initial pH on the reactive back 5 dye was studied at 4, 5, 6, 7, and 8. The results are shown in Fig. 2. The lowest residual concentration of the dye was noticed at acidic values of pH (4 and 5), while neutral and alkaline pH was not effective in the removal of the reactive dye. The explanation for the increase in the removal of the reactive dye at pH 4 and 5 is the aluminium coagulants has a high adsorption capacity for this kind of pollutants [10, 11]. The pH of 4 was used in this study for the next experiments.

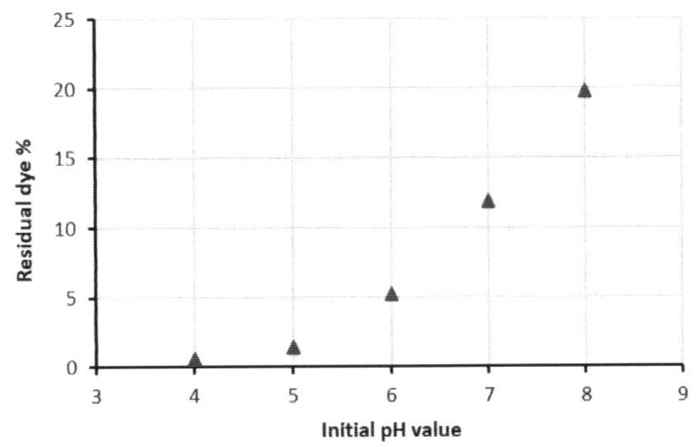

Figure 2. Effects of solution pH on dye removal.

3.2. Effects of the current density

The current density is also an important factor in the electro-coagulation process because it affects the production of aluminium coagulants from the electrodes and also the production of hydrogen bubbles. Therefore, the current density seriously influences the removal of pollutants in the electro-coagulation unit [28]. Thus, the effects of this factor were studied in this paper at three currents (current density of 1, 2 and 3 mA/cm^2).

The results are shown in Fig. 3. proved the current density strongly affects the removal of dye from water, the residual dye in the solution reduced from 5 mg/L to 1 and 0 mg/L when the current density increased from 1 to 3 mA/cm^2, respectively.

The effects of the current density are not always positive because it affects the power consumption strongly; when the current density increased from 1 to 3 mA/cm^2, the power consumption was increased by more than double the original value. Thus, 2 mA/cm^2 were used here to conduct the experiments because this value achieves good power consumption and removal of pollutants.

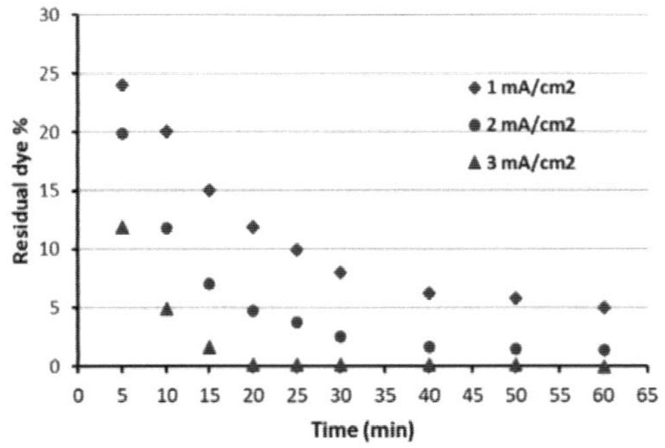

Figure 3. Effects of current density on dye removal.

It can be concluded from the results above the electro-coagulation unit can remove 82.14% of the dye when the pH, current density and treatment time were 4, 2 mA/cm^2, and 60 minutes, respectively.

The effects of the sonication process on the performance of the EC unit in terms of dye removal have been investigated by sonicating the samples for 10 minutes using a 2.7 L ultrasonic bath (FB15051) at a power of 280a constant frequency of 37000 Hz before the EC process.
The results are presented in Fig. 4. It was observed the sonication before the EC treatment increased the removal of the dye from 82.14 to 100% under the same operating conditions.
Thus, it can be concluded that the sonication process is very effective in enhancing the performance of the EC unit. This improvement is because of the ability of the ultrasonic waves to remove the accumulated oxides and sedimentations from the surface of the anodes that accelerates the production of coagulants, which helps to remove the pollutants rapidly and efficiently from water [22-25].

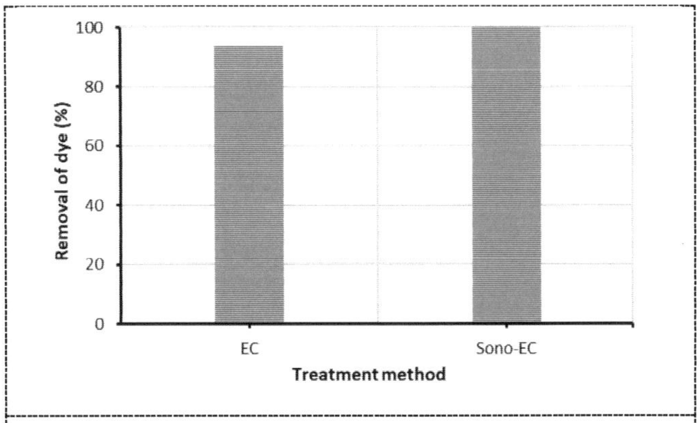

Figure 4. Effects of sonication process on the performance of the EC unit.

4. Conclusion

The present study experimentally discussed the ability of ultrasonic waves to enhance the removal of dyes from solutions using electro-coagulation technology. Electro-coagulation is known for its ability to remove many pollutants, but the accumulation of contaminants on the electrodes and the passivation of the anodes minimise the efficiency of pollutants because of the limitation of the generation of coagulants. This defect in the EC method was the focus of the presented paper. The ultrasonic waves were used to remove the accumulated pollutants and oxidation layer from the electrodes.

The presented study results showed that the electro-coagulation method removes 82.14% of the dye when the pH is low, and the current density is high. However, to achieve 100% removal of the dye, ultrasonic waves were needed. The application of the ultrasonic waves for 10 minutes improved the removal of the dye from 82.14 to 100% under the same operating conditions of the EC method.

For future researches, the effects of the frequency and power of the ultrasonic waves on the performance of the EC method must be studied.

References

[1] Alattabi A W, Harris C, Alkhaddar R, Alzeyadi A and Hashim K 2017. Treatment of Residential Complexes' Wastewater using Environmentally Friendly Technology. *Procedia Engineering,* **196** 792-9.

[2] Ryecroft S P, shaw A, Fergus P, Kot P, Hashim K and Conway L 2019. A Novel Gesomin Detection Method Based on Microwave Spectroscopy. *12th International Conference on Developments in eSystems Engineering (DeSE),* Kazan, Russia 429-33.

[3] Zubaidi S L, Al-Bugharbee H, Muhsen Y R, Hashim K, Alkhaddar R M, Al-Jumeily D and Aljaaf A J 2019. The Prediction of Municipal Water Demand in Iraq: A Case Study of Baghdad Governorate. *12th International Conference on Developments in eSystems Engineering (DeSE),* Kazan, Russia 274-7.

[4] Abdulredha M, Al Khaddar R, Jordan D, Kot P, Abdulridha A and Hashim K 2018. Estimating solid waste generation by hospitality industry during major festivals: A quantification model based on multiple regression. *Waste Management*, **77** 388-400.
[5] Alhendal M, Nasir M J, Hashim K S, Amoako-Attah J, Al-Faluji D, Muradov M, Kot P and Abdulhadi B 2020. Cost-effective hybrid filter for remediation of water from fluoride. *IOP Conference Series: Materials Science and Engineering,* 012038.
[6] Alenezi A K, Hasan H A, Hashim K S, Amoako-Attah J, Gkantou M, Muradov M, Kot P and Abdulhadi B 2020. Zeolite-assisted electrocoagulation for remediation of phosphate from calcium-phosphate solution. *IOP Conference Series: Materials Science and Engineering,* 012031.
[7] Omran I I, Al-Saati N H, Al-Saati H H, Hashim K S and Al-Saati Z N 2021. Sustainability assessment of wastewater treatment techniques in urban areas of Iraq using multi-criteria decision analysis (MCDA). *Water Practice and Technology*, **16** 648-60.
[8] Hashim K S, Shaw A, AlKhaddar R, Kot P and Al-Shamma' a A 2021. Water purification from metal ions in the presence of organic matter using electromagnetic radiation-assisted treatment. *Journal of Cleaner Production*, **280** 1-17.
[9] Aqeel K, Mubarak H A, Amoako-Attah J, Abdul-Rahaim L A, Al Khaddar R, Abdellatif M, Al-Janabi A and Hashim K S 2020. Electrochemical removal of brilliant green dye from wastewater. *IOP Conference Series: Materials Science and Engineering,* 012036.
[10] Hashim K S, Hussein A H, Zubaidi S L, Kot P, Kraidi L, Alkhaddar R, Shaw A and Alwash R 2019. Effect of initial pH value on the removal of reactive black dye from water by electrocoagulation (EC) method. *2nd International Scientific Conference,* Al-Qadisiyah University, Iraq 12-22.
[11] Hashim K S, Al-Saati N H, Alquzweeni S S, Zubaidi S L, Kot P, Kraidi L, Hussein A H, Alkhaddar R, Shaw A and Alwash R 2019. Decolourization of dye solutions by electrocoagulation: an investigation of the effect of operational parameters. *First International Conference on Civil and Environmental Engineering Technologies (ICCEET),* University of Kufa, Iraq 25-32.
[12] Abdulraheem F S, Al-Khafaji Z S, Hashim K S, Muradov M, Kot P and Shubbar A A 2020. Natural filtration unit for removal of heavy metals from water. *IOP Conference Series: Materials Science and Engineering,* 012034.
[13] Hashim K S, AlKhaddar R, Shaw A, Kot P, Al-Jumeily D, Alwash R and Aljefery M H 2020 *Electrocoagulation as an eco-friendly River water treatment method. In Advances in Water Resources Engineering and Management,* Berline: Springer.
[14] Hashim K S, Khaddar R A, Jasim N, Shaw A, Phipps D, Kot P, Pedrola M O, Alattabi A W, Abdulredha M and Alawsh R 2019. Electrocoagulation as a green technology for phosphate removal from River water. *Separation and Purification Technology*, **210** 135-44.
[15] Mohammed A-H, Hussein A H, Yeboah D, Al Khaddar R, Abdulhadi B, Shubbar A A and Hashim K S 2020. Electrochemical removal of nitrate from wastewater. *IOP Conference Series: Materials Science and Engineering,* 012037.
[16] Abdulla G, Kareem M M, Hashim K S, Muradov M, Kot P, Mubarak H A, Abdellatif M and Abdulhadi B 2020. Removal of iron from wastewater using a hybrid filter. *IOP Conference Series: Materials Science and Engineering,* 012035.
[17] Omran I I, Al-Saati N H, Hashim K S, Al-Saati Z N, Patryk K, Khaddar R A, Al-Jumeily D, Shaw A, Ruddock F and Aljefery M 2019. Assessment of heavy metal pollution in the Great Al-Mussaib irrigation channel. *Desalination and Water Treatment*, **168** 165-74.
[18] Emamjomeh M M, Mousazadeh M, Mokhtari N, Jamali H A, Makkiabadi M, Naghdali Z, Hashim K S and Ghanbari R 2020. Simultaneous removal of phenol and linear alkylbenzene sulfonate from automotive service station wastewater: Optimization of coupled electrochemical and physical processes. *Separation Science and Technology*, **55** 3184-94.

[19] Al-Marri S, AlQuzweeni S S, Hashim K S, AlKhaddar R, Kot P, AlKizwini R S, Zubaidi S L and Al-Khafaji Z S 2020. Ultrasonic-Electrocoagulation method for nitrate removal from water. *IOP Conference Series: Materials Science and Engineering,* 012073.
[20] Alyafei A, AlKizwini R S, Hashim K S, Yeboah D, Gkantou M, Al Khaddar R, Al-Faluji D and Zubaidi S L 2020. Treatment of effluents of construction industry using a combined filtration-electrocoagulation method. *IOP Conference Series: Materials Science and Engineering,* 012032.
[21] Zanki A K, Mohammad F H, Hashim K S, Muradov M, Kot P, Kareem M M and Abdulhadi B 2020. Removal of organic matter from water using ultrasonic-assisted electrocoagulation method. *IOP Conference Series: Materials Science and Engineering,* 012033.
[22] Hashim K S, Ali S S M, AlRifaie J K, Kot P, Shaw A, Al Khaddar R, Idowu I and Gkantou M 2020. Escherichia coli inactivation using a hybrid ultrasonic-electrocoagulation reactor. *Chemosphere,* **247** 125868-75.
[23] Hashim K S, Shaw A, Al Khaddar R, Ortoneda Pedrola M and Phipps D 2017. Defluoridation of drinking water using a new flow column-electrocoagulation reactor (FCER) - Experimental, statistical, and economic approach. *Journal of Environmental Management,* **197** 80-8.
[24] Alnaimi H, Idan I J, Al-Janabi A, Hashim K, Gkantou M, Zubaidi S L, Kot P and Muradov M 2020. Ultrasonic-electrochemical treatment for effluents of concrete plants. **888** 1-9.
[25] Abdulhadi B, Kot P, Hashim K, Shaw A, Muradov M and Al-Khaddar R 2021. Continuous-flow electrocoagulation (EC) process for iron removal from water: Experimental, statistical and economic study. *Science of The Total Environment,* **760** 1-16.
[26] Hashim K S, AlKhaddar R, Shaw A, Kot P, Al-Jumeily D, Alwash R and Aljefery M H 2020 *Advances in Water Resources Engineering and Management*: Springer, pp 219-35.
[27] Hashim K, Kot P, Zubaid S, Alwash R, Al Khaddar R, Shaw A, Al-Jumeily D and Aljefery M 2020. Energy efficient electrocoagulation using baffle-plates electrodes for efficient Escherichia Coli removal from Wastewater. *Journal of Water Process Engineering,* **33** 101079-86.
[28] Abdulhadi B A, Kot P, Hashim K S, Shaw A and Khaddar R A 2019. Influence of current density and electrodes spacing on reactive red 120 dye removal from dyed water using electrocoagulation/electroflotation (EC/EF) process. *First International Conference on Civil and Environmental Engineering Technologies (ICCEET),* University of Kufa, Iraq 12-22.
[29] Hashim K S, Idowu I A, Jasim N, Al Khaddar R, Shaw A, Phipps D, Kot P, Pedrola M O, Alattabi A W and Abdulredha M 2018. Removal of phosphate from River water using a new baffle plates electrochemical reactor. *MethodsX,* **5** 1413-8.
[30] Hashim K S, Shaw A, Al Khaddar R, Pedrola M O and Phipps D 2017. Energy efficient electrocoagulation using a new flow column reactor to remove nitrate from drinking water - Experimental, statistical, and economic approach. *Journal of Environmental Management,* **196** 224-33.
[31] Hashim K S, Shaw A, Al Khaddar R, Pedrola M O and Phipps D 2017. Iron removal, energy consumption and operating cost of electrocoagulation of drinking water using a new flow column reactor. *Journal of Environmental Management,* **189** 98-108.
[32] Hashim K, Shaw A and Phipps D, 2019. *Treatment reactor and method of treating a liquid.* WIPO, PCT/GB2019/052493, L J M University, United Kingdom.
[33] Hashim K S 2017 The innovative use of electrocoagulation-microwave techniques for the removal of pollutants from water. In: *Civil Engineering* (Liverpool, United Kingdom: Liverpool John Moores University) p 229

AUTHOR INDEX